ELECTRICAL PAL

2005 CODE

Paul Rosenberg

The Professional's Choice!

Electrical Pal is Volume One in the Pal®
Series of trade reference books designed
and manufactured for contractors,
maintenance personnel, service technicians
and engineers.

PAL
publications™

Pottstown, Pennsylvania
www.palpublications.com

S0-ABB-198

THIS BOOK BELONGS TO:

Name:_____

Company: _____

Title: _____

Department: _____

Company Address: _____

_____

_____

Company Phone: _____

Home Phone: _____

Date:_____

Pal Publications
374 Circle of Progress
Pottstown, PA 19464-3810

Copyright © 1996, 2000, 2002, 2005 by Pal Publications
Fifth edition published 2004

NOTICE OF RIGHTS

NOTICE OF LIABILITY

Information contained in this work has been obtained from sources believed to be reliable. However, neither Pal Publications nor its authors guarantee the accuracy or completeness of any information published herein, and neither Pal Publications nor its authors shall be responsible for any errors, omissions, damages, liabilities or personal injuries arising out of use of this information. This work is published with the understanding that Pal Publications and its authors are supplying information but are not attempting to render engineering or other professional services. If such services are required, the assistance of an appropriate professional should be sought. The reader is expressly warned to consider and adopt all safety precautions and to avoid all potential hazards. The publisher makes no representation or warranties of any kind, nor are any such representations implied with respect to the material set forth here. The publisher shall not be liable for any special, consequential, or exemplary damages resulting, in whole or part, from the readers' use of, or reliance upon, this material.

ISBN 0-9759709-0-9

08 07 06 05 04   5 4 3 2 1

Printed in the United States of America

A Note To Our Customers

We have manufactured this book to the highest quality standards possible. The cover is made of KIVAR®, a flexible, durable and water-resistant material able to withstand the toughest on-the-job conditions. We also utilize the Otabind process which allows this book to lay flatter than traditional paperback books that tend to snap shut while in use.

Other Titles From Pal Publications

Construction Pal

Data Communications Pal

Electrical Estimating Pal

Electric Motor Pal

HVAC Pal

Lighting and Maintenance Pal

Spanish Electrical Pal

Wiring Diagrams Pal

PREFACE

Many years ago I was asked to create a compendium of tables, charts, graphs, diagrams and terminology relating to the field of electricity. The Electrical Pal was my first endeavor to create a pocket guide for the electrical marketplace that would serve as an all encompassing publication in a new user-friendly format. I ended up utilizing a design and typeface that holds the most information per page while at the same time making a particular topic quick to reference.

Information covered in this manual is necessary for anyone in the field to have in one's possession at all times. Naturally, a topic may have been overlooked or not discussed in depth to suit all tradespeople. I will constantly monitor and update this book on a regular basis to not only include requested additional material, but to add new material from the ever growing amount of high technology as it develops.

Best wishes,
Paul Rosenberg

CONTENTS

Chapter 3 – *Raceways and Wiring* 3-1

Chapter 4 – *Communications* 4-1

CHAPTER 1
ELECTRICAL FORMULAS

OHM'S LAW/POWER FORMULAS

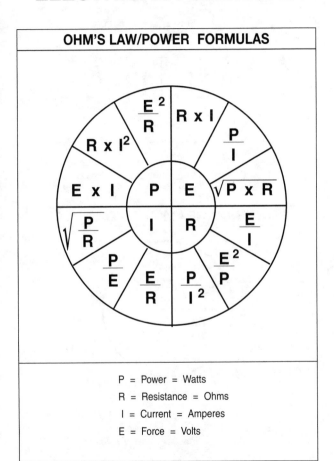

P = Power = Watts

R = Resistance = Ohms

I = Current = Amperes

E = Force = Volts

OHM'S LAW DIAGRAM AND FORMULAS

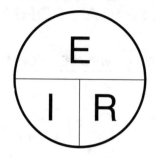

$$E = I \times R \qquad \text{Voltage} = \text{Current} \times \text{Resistance}$$
$$I = E \div R \qquad \text{Current} = \text{Voltage} \div \text{Resistance}$$
$$R = E \div I \qquad \text{Resistance} = \text{Voltage} \div \text{Current}$$

POWER DIAGRAM AND FORMULAS

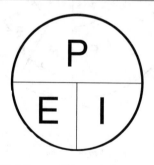

$$I = P \div E \qquad \text{Current} = \text{Power} \div \text{Voltage}$$
$$E = P \div I \qquad \text{Voltage} = \text{Power} \div \text{Current}$$
$$P = I \times E \qquad \text{Power} = \text{Current} \times \text{Voltage}$$

OHM'S LAW AND IMPEDANCE

E = VOLTAGE (IN V)
I = CURRENT (IN A)
Z = IMPEDANCE (IN Ω)

$$E = I \times Z \qquad I = \frac{E}{Z} \qquad Z = \frac{E}{I}$$

Ohm's law and the power formula are limited to circuits in which electrical resistance is the only significant opposition to current flow including all DC circuits and AC circuits that do not contain a significant amount of inductance and/or capacitance. AC circuits that include inductance are any circuits that include a coil as the load such as motors, transformers, and solenoids. AC circuits that include capacitance are any circuits that include a capacitor(s).

In DC and AC circuits that do not contain a significant amount of inductance and /or capacitance, the opposition to current flow is resistance (R). In circuits that contain inductance (X_L) or capacitance (X_C), the opposition to the flow of current is reactance (X). In circuits that contain resistance (R) and reactance (X), the combined opposition to the flow of current is impedance (Z). Resistance and Impedance are both measured in Ohms.

Ohm's law is used in circuits that contain impedance, however, and Z is substituted for R in the formula. Z represents the total resistive force (resistance and reactance) opposing current flow.

OHM'S LAW FOR ALTERNATING CURRENT

For the following Ohms Law formulas for AC current, θ is the phase angle in degrees where current lags voltage (in inductive circuit) or by which current leads voltage (in a capacitive circuit). In a resonant circuit (such as 120VAC) the phase angle is 0° and Impedance = Resistance

$$\text{Current in amps} = \frac{\text{Voltage in volts}}{\text{Impedance in ohms}}$$

$$\text{Current in amps} = \sqrt{\frac{\text{Power in watts}}{\text{Impedance in ohms} \times \cos\theta}}$$

$$\text{Current in amps} = \frac{\text{Power in watts}}{\text{Voltage in volts} \times \cos\theta}$$

$$\text{Voltage in volts} = \text{Current in amps} \times \text{Impedance in ohms}$$

$$\text{Voltage in volts} = \frac{\text{Power in watts}}{\text{Current in amps} \times \cos\theta}$$

$$\text{Voltage in volts} = \sqrt{\frac{\text{Power in watts} \times \text{Impedance in ohms}}{\cos\theta}}$$

$$\text{Impedance in ohms} = \text{Voltage in volts}/\text{Current in amps}$$

$$\text{Impedance in ohms} = \text{Power in watts}/(\text{Current in amps}^2 \times \cos\theta)$$

$$\text{Impedance in ohms} = (\text{Voltage in volts}^2 \times \cos\theta)/\text{Power in watts}$$

$$\text{Power in watts} = \text{Current in amps}^2 \times \text{Impedance in ohms} \times \cos\theta$$

$$\text{Power in watts} = \text{Current in amps} \times \text{Voltage in volts} \times \cos\theta$$

$$\text{Power in watts} = \frac{(\text{Voltage in volts})^2 \times \cos\theta}{\text{Impedance in ohms}}$$

OHM'S LAW FOR DIRECT CURRENT

Current in amps $= \dfrac{\text{Voltage in volts}}{\text{Resistance in ohms}} = \dfrac{\text{Power in watts}}{\text{Voltage in volts}}$

Current in amps $= \sqrt{\dfrac{\text{Power in watts}}{\text{Resistance in ohms}}}$

Voltage in volts = Current in amps x Resistance in ohms

Voltage in volts = Power in watts/Current in amps

Voltage in volts $= \sqrt{\text{Power in watts} \times \text{Resistance in ohms}}$

Power in watts = (Current in amps)^2 x Resistance in ohms

Power in watts = Voltage in volts x Current in amps

Power in watts = (Voltage in volts)^2/Resistance in ohms

Resistance in ohms = Voltage in volts/Current in amps

Resistance in ohms = Power in watts/(Current in amps)^2

POWER FACTOR

An AC electrical system carries two types of power:
(1) true power, watts, that pulls the load (Note: Mechanical load reflects back into an AC system as resistance.) and (2) reactive power, vars, that generates magnetism within inductive equipment. The vector sum of these two will give actual volt-amperes flowing in the circuit (see diagram right). Power factor is the cosine of the angle between true power and volt-amperes.

Volt - amperes

Reactive power / volts

Power - factor angle

True power, watts

SINGLE-PHASE POWER

Power of a single-phase AC circuit equals voltage times current times power factor:

$P_{watts} = E_{volts} \times I_{amps} \times PF.$

To figure reactive power, vars squared equals volt-amperes squared minus power squared, or

$VARS = \sqrt{(VA)^2 - (P)^2}$

THREE-PHASE AC CIRCUITS AND THE UTILIZATION OF POWER

Sine waves are actually an oscillograph trace taken at any point in a three-phase system. (Each voltage or current wave actually comes from a separate wire but are shown for comparison on common base). There are 120° between each voltage. At any instant the algebraic sum (measured up and down from centerline) of these three voltages is zero. When one voltage is zero, the other two are 86.6% maximum and have opposite signs.

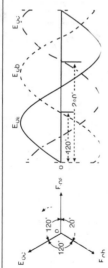

FOUR-WIRE SYSTEM

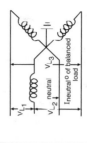

Most popular secondary distribution setup. V_1 is usually 208 V which feeds small power loads. Lighting loads at 120 V tap from any line to neutral.

DELTA CONNECTION

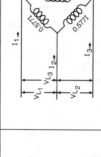

Winding voltages equal line voltages, but currents split up so 0.577 I_{line} flows through windings.

WYE-CONNECTION

Consider the three windings as primary of transformer. Current in all windings equals line current, but volts across windings = 0.577 x line volts.

TYPES OF POWER

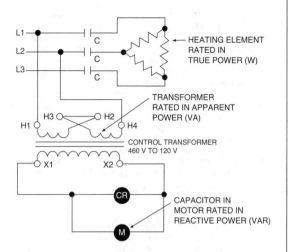

L1

L2

L3

C
C
C

HEATING ELEMENT
RATED IN
TRUE POWER (W)

TRANSFORMER
RATED IN APPARENT
POWER (VA)

H3 — H2

H1 — H4

CONTROL TRANSFORMER
460 V TO 120 V

X1 — X2

CR

M

CAPACITOR IN
MOTOR RATED IN
REACTIVE POWER (VAR)

Reactive power is supplied to a reactive load (capacitor/coil)
and is measured in volt-amps reactive (VAR). The capacitor on
a motor uses reactive power to keep the capacitor charged. The
capacitor uses no true power because it performs no actual work
such as producing heat or motion.

In an AC circuit containing only resistance, the power in
the circuit is true power. However, almost all AC circuits include
capacitive reactance (capacitors) and/or inductive reactance
(coils). Inductive reactance is the most common, because all
motors, transformers, solenoids, and coils have inductive
reactance.

Apparent power represents a load or circuit that includes
both true power and reactive power and is expressed in
volt-amps (VA), kilovolt-amps (kVA) or megavolt-amps (MVA)
Apparent power is a measure of component or system capacity
because apparent power considers circuit current regardless
of how it is used. For this reason, transformers are sized in
volt-amps rather than in watts.

TRUE POWER AND APPARENT POWER

True power is the actual power used in an electrical circuit and is expressed in watts (W). *Apparent power* is the product of voltage and current in a circuit calculated without considering the phase shift that may be present between total voltage and current in the circuit. Apparent power is measured in volt-amperes (VA). A phase shift exists in most AC circuits that contain devices causing capacitance or inductance.

True power equals apparent power in an electrical circuit containing only resistance. True power is less than apparent power in a circuit containing inductance or capacitance.

Capacitance is the property of a device that permits the storage of electrically separated charges when potential differences exist between the conductors. *Inductance* is the property of a circuit that causes it to oppose a change in current due to energy stored in a magnetic field; i.e., coils.

To calculate true power, apply the formula:

$$P_T = I^2 \times R$$

where

P_T = true power (in watts)

I = total circuit current (in amps)

R = total resistive component of the circuit (in ohms)

To calculate apparent power, apply the formula:

$$P_A = E \times I$$

where

P_A = apparent power (in volt-amps)

E = measured voltage (in volts)

I = measured current (in amps)

POWER FACTOR FORMULA

Power factor is the ratio of true power used in an AC circuit to apparent power delivered to the circuit.

$$PF = \frac{P_T}{P_A}$$

where

PF = power factor (percentage)

P_T = true power (in watts)

P_A = apparent power (in volt-amps)

POWER FACTOR IMPROVEMENT
Capacitor Multipliers for Kilowatt Load

To give capacitor kvar required to improve power factor from original to desired value. For example, assume the total plant load is 100kw at 60 percent power factor. Capacitor kvar rating necessary to improve power factor to 80 percent is found by multiplying kw (100) by multiplier in table (0.583), which gives kvar (58.3). Nearest standard rating (60 kvar) should be recommended.

Original Power Factor Percentage	Desired Power Factor Percentage				
	100%	95%	90%	85%	80%
60	1.333	1.004	0.849	0.713	0.583
62	1.266	0.937	0.782	0.646	0.516
64	1.201	0.872	0.717	0.581	0.451
66	1.138	0.809	0.654	0.518	0.388
68	1.078	0.749	0.594	0.458	0.338
70	1.020	0.691	0.536	0.400	0.270
72	0.964	0.635	0.480	0.344	0.214
74	0.909	0.580	0.425	0.289	0.159
76	0.855	0.526	0.371	0.235	0.105
77	0.829	0.500	0.345	0.209	0.079
78	0.802	0.473	0.318	0.182	0.052
79	0.776	0.447	0.292	0.156	0.026
80	0.750	0.421	0.266	0.130	
81	0.724	0.395	0.240	0.104	
82	0.698	0.369	0.214	0.078	
83	0.672	0.343	0.188	0.052	
84	0.646	0.317	0.162	0.206	
85	0.620	0.291	0.136		
86	0.593	0.264	0.109		
87	0.567	0.238	0.083		
88	0.540	0.211	0.056		
89	0.512	0.183	0.028		
90	0.484	0.155			
91	0.456	0.127			
92	0.426	0.097			
93	0.395	0.066			
94	0.363	0.034			
95	0.329				
96	0.292				
97	0.251				
99	0.143				

VOLT AND POWER RATIOS VS. DECIBELS

Voltage	Power	+ DB -	Power	Voltage
1.000	1.000	0.0	1.000	1.000
1.059	1.122	0.5	0.891	0.944
1.122	1.259	1.0	0.794	0.891
1.189	1.413	1.5	0.708	0.847
1.259	1.585	2.0	0.631	0.794
1.334	1.778	2.5	0.562	0.750
1.413	1.995	3.0	0.501	0.708
1.496	2.239	3.5	0.447	0.668
1.585	2.512	4.0	0.398	0.631
1.679	2.818	4.5	0.355	0.596
1.778	3.162	5.0	0.316	0.562
1.884	3.548	5.5	0.282	0.531
1.995	3.981	6.0	0.251	0.501
2.113	4.467	6.5	0.224	0.473
2.239	5.012	7.0	0.200	0.447
2.371	5.623	7.5	0.178	0.422
2.512	6.310	8.0	0.158	0.398
2.661	7.079	8.5	0.141	0.376
2.818	7.943	9.0	0.126	0.355
2.985	8.913	9.5	0.112	0.335
3.162	10.000	10.0	0.100	0.316
3.350	11.220	10.5	0.089	0.299
3.548	12.589	11.0	0.079	0.282
3.758	14.125	11.5	0.071	0.266
3.981	15.849	12.0	0.063	0.251
4.217	17.783	12.5	0.056	0.237
4.467	19.953	13.0	0.050	0.224
4.732	22.387	13.5	0.045	0.211
5.012	25.119	14.0	0.040	0.200
5.309	28.184	14.5	0.035	0.188
5.623	31.623	15.0	0.032	0.178
5.957	35.481	15.5	0.028	0.168
6.310	39.811	16.0	0.025	0.158
6.683	44.668	16.5	0.022	0.150
7.079	50.119	17.0	0.020	0.141
7.499	56.234	17.5	0.018	0.133
7.943	63.096	18.0	0.016	0.126
8.414	70.795	18.5	0.014	0.119
8.913	79.433	19.0	0.013	0.112
9.441	89.125	19.5	0.011	0.106
10.0	100	20.0	0.010	0.100
31.6	1000	30.0	0.001	0.0316
100.0	10000	40.0	0.0001	0.01
316.2	100000	50.0	0.00001	0.00316
1000	10^6	60.0	10^{-6}	0.001
3162	10^7	70.0	10^{-7}	0.000316
10000	10^8	80.0	10^{-8}	0.0001
31620	10^9	90.0	10^{-9}	0.0000316
100000	10^{10}	100.0	10^{-10}	0.00001
316200	10^{11}	110.0	10^{-11}	0.00000316
1000000	10^{12}	120.0	10^{-12}	0.000001

AC/DC POWER FORMULAS

TO FIND	FOR DIRECT CURRENT	FOR ALTERNATING CURRENT		
		1φ, 115 or 120 V	1φ, 208, 230 or 240 V	3φ - ALL VOLTAGES
AMPERES WHEN HORSEPOWER IS KNOWN	$\dfrac{HP \times 746}{E \times E_{FF}}$	$\dfrac{HP \times 746}{E \times E_{FF} \times PF}$	$\dfrac{HP \times 746}{E \times E_{FF} \times PF}$	$\dfrac{HP \times 746}{1.73 \times E \times E_{FF} \times PF}$
AMPERES WHEN KILOWATTS IS KNOWN	$\dfrac{kW \times 1000}{E}$	$\dfrac{kW \times 1000}{E \times PF}$	$\dfrac{kW \times 1000}{E \times PF}$	$\dfrac{kW \times 1000}{1.73 \times E \times PF}$
AMPERES WHEN kVA IS KNOWN		$\dfrac{kVA \times 1000}{E}$	$\dfrac{kVA \times 1000}{E}$	$\dfrac{kVA \times 1000}{1.73 \times E}$
KILOWATTS	$\dfrac{I \times E}{1000}$	$\dfrac{I \times E \times PF}{1000}$	$\dfrac{I \times E \times PF}{1000}$	$\dfrac{I \times E \times 1.73 \times PF}{1000}$
KILOVOLT-AMPS		$\dfrac{I \times E}{1000}$	$\dfrac{I \times E}{1000}$	$\dfrac{I \times E \times 1.73}{1000}$
HORSEPOWER (OUTPUT)	$\dfrac{I \times E \times E_{FF}}{746}$	$\dfrac{I \times E \times E_{FF} \times PF}{746}$	$\dfrac{I \times E \times E_{FF} \times PF}{746}$	$\dfrac{I \times E \times 1.73 \times E_{FF} \times PF}{746}$

HORSEPOWER FORMULAS

Current and Voltage Known

$$HP = \frac{E \times I \times E_{ff}}{746}$$

where

HP = horsepower
I = current (amps)
E = voltage (volts)
E_{ff} = efficiency
746 = constant

Speed and Torque Known

$$HP = \frac{rpm \times T}{5252}$$

where

HP = horsepower
rpm = revolutions per minute
T = torque (lb-ft)
5252 = constant

EFFICIENCY FORMULAS

Input and Output Power Known

$$E_{ff} = \frac{P_{out}}{P_{in}}$$

where

E_{ff} = efficiency (%)
P_{out} = output power (W)
P_{in} = input power (W)

Horsepower and Power Loss Known

$$E_{ff} = \frac{746 \times HP}{(746 \times HP) + W_l}$$

where

E_{ff} = efficiency (%)
746 = constant
HP = horsepower
W_l = watts lost

VOLTAGE UNBALANCE

$$V_u = \frac{V_d}{V_a} \times 100$$

where

V_u = voltage unbalance (%)
V_d = voltage deviation (V)
V_a = voltage average (V)
100 = constant

TEMPERATURE CONVERSIONS

Convert °C to °F

$$°F = (1.8 \times °C) + 32$$

Convert °F to °C

$$°C = \frac{(°F - 32)}{1.8}$$

VOLTAGE DROP FORMULAS

The NEC® recommends a maximum 3% voltage drop for either the branch circuit or the feeder.

Single-Phase:

$$VD = \frac{2 \times R \times I \times L}{CM}$$

Three-Phase:

$$VD = \frac{1.732 \times R \times I \times L}{CM}$$

VD = Volts (voltage drop of the circuit)

R = 12.9 Ohms/Copper or 21.2 Ohms/Aluminum (resistance constants for a 1,000 circular mils conductor that is 1,000 feet long, at an operating temperature of 75° C.)

I = Amps (load at 100 percent)

L = Feet (length of circuit from load to power supply)

CM = Circular-Mils (conductor wire size)

2 = Single-Phase Constant

1.732 = Three-Phase Constant

CONDUCTOR LENGTH/VOLTAGE DROP

Voltage drop can be reduced by limiting the length of the conductors.

Single-Phase:

$$L = \frac{CM \times VD}{2 \times R \times I}$$

Three-Phase:

$$L = \frac{CM \times VD}{1.732 \times R \times I}$$

CONDUCTOR SIZE/VOLTAGE DROP

Increase the size of the conductor to decrease the voltage drop of circuit (reduce its resistance).

Single-Phase:

$$CM = \frac{2 \times R \times I \times L}{VD}$$

Three-Phase:

$$CM = \frac{1.732 \times R \times I \times L}{VD}$$

FORMULAS FOR SINE WAVES

Frequency	Period	Peak-to-Peak Value
$f = \dfrac{1}{T}$	$T = \dfrac{1}{f}$	$V_{p\text{-}p} = 2 \times V_{max}$
where f = frequency (in hertz) 1 = constant T = period (in seconds)	where T = period (in seconds) 1 = constant f = frequency (in hertz)	where $V_{p\text{-}p}$ = peak-to-peak value 2 = constant V_{max} = peak value

Average Value	RMS Value
$V_{avg} = V_{max} \times .637$	$V_{rms} = V_{max} \times .707$
where V_{avg} = average value (in volts) V_{max} = peak value (in volts) .637 = constant	where V_{rms} = rms value (in volts) V_{max} = peak value (in volts) .707 = constant

CALCULATING ROOT-MEAN-SQUARE (RMS)

Effective (RMS) value = 0.707 x Peak value
Effective (RMS) value = 1.11 x Average value
Average value = 0.637 x Peak value
Average value = 0.9 x Effective (RMS) value
Peak value = 1.414 x Effective (RMS) value
Peak value = 1.57 x Average value
Peak-to-Peak value = 2 x Peak value
Peak-to-Peak value = 2.828 x Effective (RMS) value

LOCKED ROTOR CURRENT

Apparent, 1φ

$$LRC = \frac{1000 \times HP \times kVA/HP}{V}$$

Apparent, 3φ

$$LRC = \frac{1000 \times HP \times kVA/HP}{V \times \sqrt{3}}$$

True, 1φ

$$LRC = \frac{1000 \times HP \times kVA/HP}{V \times PF \times E_{ff}}$$

True, 3φ

$$LRC = \frac{1000 \times HP \times kVA/HP}{V \times \sqrt{3} \times PF \times E_{ff}}$$

where

LRC = locked rotor current (in amps)

1000 = multiplier for kilo

HP = horsepower

kVA/HP = kilovolt-amps per horsepower

V = volts

$\sqrt{3}$ = 1.732

PF = power factor

E_{ff} = motor efficiency

MAXIMUM OCPD

$OCPD = FLC \times R_M$

where

FLC = full load current (from motor nameplate or NEC® Table 430.150)

R_M = maximum rating of OCPD

Motor Type	Code Letter	Motor Size	FLC %			
			TDF	NTDF	ITB	ITCB
AC*	—	—	175	300	150	700
AC*	A	—	150	150	150	700
AC*	B—E	—	175	250	200	700
AC*	F—V	—	175	300	250	700
DC	—	1/8 TO 50 HP	150	150	150	250
DC	—	Over 50 HP	150	150	150	175

* full-voltage and resistor starting

MOTOR TORQUE FORMULAS

Torque	Starting Torque	Nominal Torque Rating
$T = \dfrac{HP \times 5252}{rpm}$	$T = \dfrac{HP \times 5252 \times \%}{rpm}$	$T = \dfrac{HP \times 63,000}{rpm}$
where	where	where
T = torque	HP = horsepower	T = nominal torque rating (in lb-in)
HP = horsepower	$5252 = $ constant $\left(\dfrac{33,000 \text{ lb-ft}}{\pi \times 2} = 5252\right)$	$63,000 = $ constant
$5252 = $ constant $\left(\dfrac{33,000 \text{ lb-ft}}{\pi \times 2} = 5252\right)$	rpm = revolutions per minute	HP = horsepower
rpm = revolutions per minute	$\%$ = motor class percentage	rpm = revolutions per minute

GEAR REDUCER FORMULAS

Output Torque	Output Speed	Output Horsepower
$O_T = I_T \times R_R \times R_E$	$O_S = \dfrac{I_S}{R_R} \times R_E$	$O_{HP} = I_{HP} \times R_E$
where	where	where
O_T = output torque (in lb.-ft.)	O_S = output speed (in rpm)	O_{HP} = output horsepower
I_T = input torque (in lb.-ft.)	I_S = input speed (in rpm)	I_{HP} = input horsepower
R_R = gear reducer ratio	R_R = gear reducer ratio	R_E = reducer efficiency %
R_E = reducer efficiency %	R_E = reducer efficiency %	

SUMMARY OF SERIES, PARALLEL, AND COMBINATION CIRCUITS

To Find	Series Circuits	Parallel Circuits	Series/Parallel
Resistance (R) Ohm Ω	$R_T = R_1 + R_2 + R_3$ Sum of individual resistances	$\dfrac{1}{R_T} = \dfrac{1}{R_1} + \dfrac{1}{R_2} + \dfrac{1}{R_3}$	Total resistance equals resistance of parallel portion and sum of series resistors
Current (I) Ampere A	$I_T = I_1 = I_2 = I_3$ The same throughout entire circuit	$I_T = I_1 + I_2 + I_3$ Sum of individual currents	Series rules apply to series portion of circuit Parallel rules apply to parallel part of circuit
Voltage (E) Volt V, E	$E_T = E_1 + E_2 + E_3$ Sum of individual voltages	$E_T = E_1 = E_2 = E_3$ Total voltage and branch voltage are the same	Total voltage is sum of voltage drops across each series resistor and each of the branches of parallel portion
Power (P) Watt W	$P_T = P_1 + P_2 + P_3$ Sum of individual wattages	$P_T = P_1 + P_2 + P_3$ Sum of Individual wattages	$P_T = P_1 + P_2 + P_3$ Sum of Individual wattages

CIRCUIT CHARACTERISTICS

Resistances in a Series DC Circuit

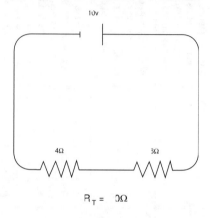

Parallel Circuit, Showing Voltage Drops

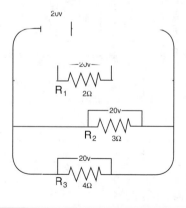

CIRCUIT CHARACTERISTICS (cont.)

Voltages in a Series-Parallel Circuit

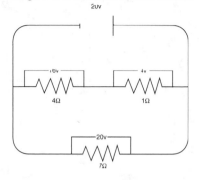

Parallel Circuit, Showing Current Values

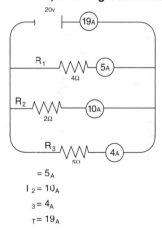

$$= 5_A$$
$$I_2 = 10_A$$
$$_3 = 4_A$$
$$_T = 19_A$$

RESISTANCE

Resistors in Series

$$R_T = R_1 + R_2 + R_3 + \cdots$$

Resistors in Parallel

$$R_T = \frac{R_1 \times R_2}{R_1 + R_2} \quad \text{(Two Only)}$$

$$\frac{1}{R_T} = \frac{1}{R_1} + \frac{1}{R_2} + \frac{1}{R_3} + \cdots \quad \text{(Multiple)}$$

RESISTORS CONNECTED IN SERIES/PARALLEL

(Two Only)

$$R_T = \left(\frac{R_1 \times R_2}{R_1 + R_2} \right) + R_1 + R_2 + R_3 + \cdots$$

(Multiple)

$$R_T = \left(\frac{1}{\frac{1}{R_1} + \frac{1}{R_2} + \frac{1}{R_3} \cdots} \right) + R_1 + R_2 + R_3 + \cdots$$

CAPACITANCE

Capacitors in Series

$$C_T = \frac{C_1 \times C_2}{C_1 + C_2} \quad \text{(Two only)}$$

$$\frac{1}{C_T} = \frac{1}{C_1} + \frac{1}{C_2} + \frac{1}{C_3} + \ldots \text{ (Multiple)}$$

Capacitors in Parallel

$$C_T = C_1 + C_2 + C_3 + \ldots$$

CAPACITORS CONNECTED IN SERIES/PARALLEL

1. Calculate the capacitance of the parallel branch.

$$C_{PT} = C_1 + C_2 + C_3 + \ldots$$

2. Calculate the capacitance of the series combination.

$$C_T = \frac{C_{PT} \times C_S}{C_{PT} + C_S}$$

INDUCTANCE

Inductors in Series

$$L_T = L_1 + L_2 + L_3 + \ldots$$

Inductors in Parallel

$$L_T = \frac{L_1 \times L_2}{L_1 + L_2} \quad \text{(Two Only)}$$

$$\frac{1}{L_T} = \frac{1}{L_1} + \frac{1}{L_2} + \frac{1}{L_3} + \ldots \quad \text{(Multiple)}$$

INDUCTORS CONNECTED IN SERIES/PARALLEL

(Two Only)

$$L_T = \left(\frac{L_1 \times L_2}{L_1 + L_2} \right) + L_1 + L_2 + L_3 + \ldots$$

(Multiple)

$$L_T = \left(\frac{1}{\frac{1}{L_1} + \frac{1}{L_2} + \frac{1}{L_3} + \ldots} \right) + L_1 + L_2 + L_3 + \ldots$$

INDUCTIVE REACTANCE

Inductive reactance (X_L) is an inductor's opposition to alternating current measured in ohms.

To calculate inductive reactance for AC circuits, apply the formula:

$$X_L = 2\pi fL$$

where

X_L = inductive reactance (in Ohms)

2π = 6.28 (indicates circular motion that produces the AC sine wave)

f = applied frequency (in Hertz)

L = inductance (in Henrys)

Inductive Reactances

In Series

$$X_{L_T} = X_{L_1} + X_{L_2} + \ldots$$

In Parallel

$$X_{LT} = \frac{X_{L_1} \times X_{L_2}}{X_{L_1} + X_{L_2}} \quad \text{(2 Only)}$$

$$\frac{1}{X_{L_T}} = \frac{1}{X_{L_1}} + \frac{1}{X_{L_2}} + \frac{1}{X_{L_3}} \ldots \quad \text{(Multiple)}$$

To calculate inductive reactance when voltage across a coil (E_L) and current through a coil (I_L) are known.

$$X_L = \frac{E_L}{I_L}$$

where

E_L = inductive reactance (in Ohms)

E_L = voltage across coil (in Volts)

I_L = current through coil (in Amps)

CAPACITIVE REACTANCE

Capacitive reactance (X_C) is the opposition to current flow by a capacitor when connected to an AC power supply and is expressed in ohms. To calculate capacitive reactance, apply the formula:

$$X_C = \frac{1}{2\pi fC}$$

where

X_C = capacitive reactance (in Ohms)

2π = 6.28 (indicates circular motion that produces the AC sine wave)

f = applied frequency (in Hertz)

C = capacitance (in Farads)

Capacitive Reactances

In Series

$$X_{C_T} = X_{C_1} + X_{C_2} + X_{C_3}...$$

In Parallel

$$X_{CT} = \frac{X_{C_1} \times X_{C_2}}{X_{C_1} + X_{C_2}} \quad \text{(2 Only)}$$

$$\frac{1}{X_{C_T}} = \frac{1}{X_{C_1}} + \frac{1}{X_{C_2}} + \frac{1}{X_{C_3}} +... \quad \text{(Multiple)}$$

To calculate capacitive reactance when voltage across the capacitor (E_C) and current through the capacitor (I_C) are known, apply the formula:

$$X_C = \frac{E_C}{I_C}$$

where

X_C = capacative reactance (in Ohms)

E_C = voltage across capacitor (in Volts)

I_C = current through capacitor (in Amps)

DELTA AND WYE RESISTOR CIRCUITS

In the delta network the resistance between terminals may be determined by combining the formulas for series and parallel resistances as follows:

$$a \text{ to } b = \frac{R_z \times (R_x + R_y)}{R_z + (R_x + R_y)}$$

$$a \text{ to } c = \frac{R_y \times (R_x + R_z)}{R_y + (R_x + R_z)}$$

$$b \text{ to } c = \frac{R_x \times (R_z + R_y)}{R_x + (R_z + R_y)}$$

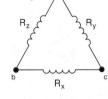

Delta

To convert from delta to wye, use:

$$R_a = \frac{R_y \times R_z}{R_x + R_y + R_z}$$

$$R_b = \frac{R_x \times R_z}{R_x + R_y + R_z}$$

Wye

$$R_c = \frac{R_x \times R_y}{R_x + R_y + R_z}$$

To convert from wye to delta, use:

$$R_x = \frac{(R_a \times R_b) + (R_b \times R_c) + (R_c \times R_a)}{R_a}$$

$$R_y = \frac{(R_a \times R_b) + (R_b \times R_c) + (R_c \times R_a)}{R_b}$$

$$R_z = \frac{(R_a \times R_b) + (R_b \times R_c) + (R_c \times R_a)}{R_c}$$

ELECTRICAL FORMULAS

Quantity of Electricity in a Capacitor:

Q in coulombs = Capacitance in farads x Volts

Q or Figure of Merit:

Q = Inductive Reactance in ohms/Series Resistance in ohms
Q = Capacitive Reactance in ohms/Series Resistance in ohms

Capacitance of a Capacitor:

Capacitance in picofarads =

$$0.0885 \times \frac{\text{Dielectric constant} \times \text{area in cm}^2 \times (\text{\# of plates - 1})}{\text{thickness of dielectric in cm}}$$

Self Inductance:

When including the effects of coupling, add 2 x mutual inductance if fields are adding and subtract 2 x mutual inductance if the fields are opposing. Examples below:

$$\text{Series: } L_t = L_1 + L_2 + 2M \quad \text{or} \quad L_t = L_1 + L_2 - 2M$$

$$\text{Parallel: } L_T = \frac{1}{[(1/L_1 + M) + (1/L_2 + M)]}$$

Conductive Leakage Current:

$$I_L = \frac{V_a}{R_1}$$

where

I_L = leakage current (in microamperes)

V_a = applied voltage (in volts)

R_1 = insulation resistance (in megohms)

ELECTRICAL FORMULAS (cont.)

Resonance: (f)

Resonant frequency in hertz (where $X_L = X_C$) =

$$\frac{1}{2\pi \sqrt{\text{Inductance in henrys x Capacitance in farads}}}$$

Reactance: (X)

Reactance in ohms of an inductance is X_L
Reactance in ohms of a capacitance is X_C

$X_L = 2\pi FH$
$X_c = 1 \div 2\pi FC$
F = frequency, in hertz, H = inductance, in henrys,
and C = capacitance, in farads.

Impedance: (Z)

Impedance in ohms = $\sqrt{\text{Resistance in ohms}_2 + (X_L - X_C)^2}$
(series)

Impedance in ohms = $\dfrac{\text{Resistance in ohms} + \text{Reactance}}{\sqrt{\text{Resistance in ohms}^2 + \text{Reactance}^2}}$
(parallel)

Susceptance: (B)

Susceptance in mhos = $\dfrac{\text{Reactance in ohms}}{\text{Resistance in ohms}^2 + \text{Reactance in ohms}^2}$

Admittance: (Y)

Admittance in mhos = $\dfrac{1}{\sqrt{\text{Resistance in ohms}^2 \text{ x Reactance in ohms}^2}}$

Admittance in mhos = 1/Impedance in ohms

ELECTRICAL FORMULAS (cont.)

Power Factor: (PF)

Power Factor = cos (Phase Angle)

Power Factor = True Power/Apparent Power

Power Factor = Power in watts/volts x current in amps

Power Factor = Resistance in ohms/Impedance in ohms

Efficiency:

Efficiency = Output/Input

Sine Wave Voltage and Current:

Effective (RMS) value = 0.707 x Peak value

Effective (RMS) value = 1.11 x Average value

Average value = 0.637 x Peak value

Average value = 0.9 x Effective (RMS) value

Peak value = 1.414 x Effective (RMS) value

Peak value = 1.57 x Average value

Decibels: (dB)

dB = 10 Log $_{10}$ (power in Watts #1/Power in Watts #2)

dB = 10 Log $_{10}$ (Power Ratio)

dB = 20 Log $_{10}$ (Volts or Amps #1/Volts or Amps #2)

dB = 20 Log $_{10}$ (Volts or Current Ratio)

Power Ratio = $10^{(dB/10)}$

Voltage or Current ratio = $10^{(dB/10)}$

If impedances are not equal: $dB = 20 \log_{10}\left[\left(Volt_1\sqrt{Z_2}\right) \div \left(Volt_2\sqrt{Z_1}\right)\right]$

ELECTRICAL FORMULAS (cont.)

Wavelength and Frequency: (λ) and (f)

Wavelength in meters = (300,000)/frequency in kilohertz
Wavelength in meters = (300)/frequency in megahertz
Wavelength in meters = (984)/frequency in megahertz
Frequency in kilohertz = (300,000)/wavelength in meters
Frequency in megahertz = (300)/wavelength in meters
Frequency in megahertz = (984)/wavelength in feet

Antenna Length:

Quarter-wave antenna:
Length in feet = 234/frequency in megahertz

Half-wave antenna:
Length in feet = 468/frequency in megahertz

LCR Series Time Circuits:

Time in seconds =
Inductance in henrys/Resistance in ohms

Time in seconds =
Capacitance in farads x Resistance in ohms

Loud Speaker Matching Transformer:

Transformer Primary Impedance =
(Amplifier output volts)^2/Speaker Power

Time Duration of One Cycle:

100 kilohertz	=	10 microsecond cycle
250 kilohertz	=	4 microsecond cycle
1 megahertz	=	1 microsecond cycle
4 megahertz	=	250 nanoseconds cycle
10 megahertz	=	100 nanoseconds cycle

TRANSFORMER TURNS RATIO

A transformer increases or decreases voltage by inducing electrical energy from one coil to another through magnetic lines of force. The turns ratio is the ratio between the voltage and the number of turns on the primary and secondary windings of a transformer.

Thus

$$\frac{N_p}{N_s} = \frac{E_p}{E_s}$$

where

N_p = number of turns in primary

N_s = number of turns in secondary

E_p = primary voltage (in volts)

E_s = secondary voltage (in volts)

TRANSFORMER INVERSE RATIO

An inverse ratio exists between the number of turns, the voltage, and the current in the primary and secondary of a transformer. To determine these inverse ratios, apply the following formulas:

$$\frac{N_p}{N_s} = \frac{I_s}{I_p} \quad \text{or} \quad \frac{E_p}{E_s} = \frac{I_s}{I_p}$$

where

N_p = number of turns in primary

N_s = number of turns in secondary

E_p = primary voltage (in volts)

E_s = secondary voltage (in volts)

I_p = primary current (in amps)

I_s = secondary current (in amps)

CHAPTER 2
ELECTRONICS AND
ELECTRONIC SYMBOLS

COLOR CODES FOR RESISTORS

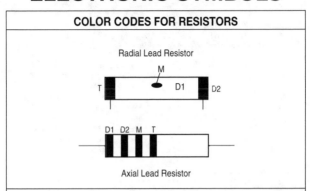

Radial Lead Resistor

Axial Lead Resistor

Color	1st # (D1)	2nd # (D2)	Multiplier (M)	Tolerance (T)
No Color				20%
Silver			0.01	10%
Gold			0.1	5%
White	9	9	10^9	
Gray	8	8	100,000,000	
Violet	7	7	10,000,000	
Blue	6	6	1,000,000	
Green	5	5	100,000	
Yellow	4	4	10,000	
Orange	3	3	1,000	4%
Red	2	2	100	3%
Brown	1	1	10	2%
Black	0	0	1	1%

Example: Yellow – Blue – Brown – Silver = 460 ohms

If only Band D1 is wide, it indicates that the resistor is wirewound. If Band D1 is wide and there is also a blue fifth band to the right of Band T on the Axial Lead Resistor, it indicates the resistor is wirewound and flame proof.

STANDARD VALUES FOR RESISTORS

Values for 5% class   k = kilohms = 1,000 ohms   m = megohms = 1,000,000 ohms

1	8.2	68	560	4.7k	39k	330k	2.7m
1.1	9.1	75	620	5.1k	43k	360k	3.0m
1.2	10	82	680	5.6k	47k	390k	3.3m
1.3	11	91	750	6.2k	51k	430k	3.6m
1.5	12	100	820	6.8k	56k	470k	3.9m
1.6	13	110	910	7.5k	62k	510k	4.3m
1.8	15	120	1.0k	8.2k	68k	560k	4.7m
2.0	16	130	1.1k	9.1k	75k	620k	5.1m
2.2	18	150	1.2k	10k	82k	680k	5.6m
2.4	20	160	1.3k	11k	91k	750k	6.2m
2.7	22	180	1.5k	12k	100k	820k	6.8m
3.0	24	200	1.6k	13k	110k	910k	7.5m
3.3	28	220	1.8k	15k	120k	1.0m	8.2m
3.6	30	240	2.0k	16k	130k	1.1m	9.1m
3.9	33	270	2.2k	18k	150k	1.2m	10.0m
4.3	36	300	2.4k	20k	160k	1.3m	
4.7	39	330	2.7k	22k	180k	1.5m	
5.1	43	360	3.0k	24k	200k	1.6m	
5.6	47	390	3.3k	27k	220k	1.8m	
6.2	51	430	3.6k	30k	240k	2.0m	
6.8	56	470	3.9k	33k	270k	2.2m	
7.5	62	510	4.3k	36k	300k	2.4m	

CAPACITORS

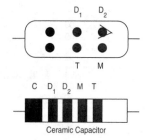

Mica Capacitor

D_1 D_2

T   M

C D_1 D_2 M   T

Ceramic Capacitor

Disc Capacitor

D_2

C D_1 M   T

Ceramic disc capacitors are normally labeled. If the number is less than 1 then the value is in picofarads, if greater than 1 the value is in microfarads. The letter R can be used in place of a decimal; for example, 2R9=2.9

COLOR CODES FOR CAPACITORS

Color	1st # (D1)	2nd # (D2)	Multiplier (M)	Tolerance (T)
No Color				20%
Silver			0.01	10%
Gold			0.1	5%
White	9	9	10^9	9%
Gray	8	8	100,000,000	8%
Violet	7	7	10,000,000	7%
Blue	6	6	1,000,000	6%
Green	5	5	100,000	5%
Yellow	4	4	10,000	4%
Orange	3	3	1,000	3%
Red	2	2	100	2%
Brown	1	1	10	1%
Black	0	0	1	20%

COLOR CODES FOR CERAMIC CAPACITORS

Color	Tolerance (T) Above 10pf	Tolerance (T) Below 10pf	Decimal Multiplier (M)	Temp Coef ppm/°C (C)
White	10	1.0	0.1	500
Gray	–	0.25	0.01	30
Violet	–	–	–	-750
Blue	–	–	–	-470
Green	5	0.5	–	-330
Yellow	–	–	–	-220
Orange	–	–	1000	-150
Red	2	–	100	-80
Brown	1	–	10	-30
Black	20	2.0	1	0

STANDARD VALUES FOR CAPACITORS

pF	mF	mF	mF	mF
10	0.0010	0.10	10	1000
12	0.0012			
13	0.0013			
15	0.0015	0.15	15	
18	0.0018			
20	0.0020			
22	0.0022	0.22	22	2200
24				
27				
30				
33	0.0033	0.33	33	3300
36				
43				
47	0.0047	0.47	47	4700
51				
56				
62				
68	0.0068	0.68	68	6800
75				
82				
100	0.0100	1.00	100	10000
110				
120				
130				
150	0.0150	1.50		
180				
200				
220	0.0220	2.20	220	22000
240				
270				
300				
330	0.0330	3.30	330	
360				
390				
430				
470	0.0470	4.70	470	47000
510				
560				
620				
680	0.0680	6.80		
750				
820				82000
910				

pF = picofarads = 1 x 10^{-12} farads   mF = microfarads = 1 x 10^{-6} farads

CAPACITOR RATINGS

110-125 VAC, 50/60 Hz, Starting Capacitors

Typical Ratings*	Dimensions**		Model Number***
	Diameter	Length	
88-106	1-7/16	2-3/4	EC8815
108-130	1-7/16	2-3/4	EC10815
130-489	1-7/16	2-3/4	EC13015
145-174	1-7/16	2-3/4	EC14515
161-193	1-7/16	2-3/4	EC16115
189-227	1-7/16	2-3/4	EC18915A
216-259	1-7/16	3-3/8	EC21615
233-280	1-7/16	3-3/8	EC23315A
243-292	1-7/16	3-3/8	EC24315A
270-324	1-7/16	3-3/8	EC27015A
324-389	1-7/16	3-3/8	EC2R10324N
340-408	1-13/16	3-3/8	EC34015
378-454	1-13/16	3-3/8	EC37815
400-480	1-13/16	3-3/8	EC40015
430-516	1-13/16	3-3/8	EC43015A
460-553	1-13/16	4-3/8	EC5R10460N
540-648	1-13/16	4-3/8	EC54015B
590-708	1-13/16	4-3/8	EC59015A
708-850	1-13/16	4-3/8	EC70815
815-978	1-13/16	4-3/8	EC81515
1000-1200	2-1/16	4-3/8	EC100015A

220-250 VAC, 50/60 Hz, Starting Capacitors

53-64	1-7/16	3-3/8	EC5335
64-77	1-7/16	3-3/8	EC6435
88-406	1-13/16	3-3/8	EC8835
108-130	1-13/16	3-3/8	EC10835A
124-149	1-13/16	4-3/8	EC12435
130-154	1-13/16	4-3/8	EC13035
145-174	2-1/16	3-3/8	EC6R22145N
161-193	2-1/16	3-3/8	EC6R2216N
216-259	2-1/16	4-3/8	EC21635A
233-280	2-1/16	4-3/8	EC23335A
270-324	2-1/16	4-3/8	EC27035A

* in µF
** in inches
*** Model numbers vary by manufacturer

CAPACITOR RATINGS (cont.)

270 VAC, 50/60 Hz, Running Capacitors

Typical Ratings*	Dimensions**		Model Number***
	Oval	Length	
2		2-1/8	VH550
3		2-1/8	VH25503
4	1-5/16 x 2-5/32	2-1/8	VH5704
5		2-1/8	VH5705
6		2-5/8	VH5706
7.5		2-7/8	VH9001
10	1-5/16 x 2-5/32	2-7/8	VH9002
12.5		3-7/8	VH9003
15	1-29/32 x 2-29/32	3-1/8	VH9121
17.5		2-7/8	VH9123
20		2-7/8	VH5463
25	1-29/32 x 2-29/32	3-7/8	VH9069
30		3-7/8	VH5465
35	1-29/32 x 2-29/32	3-7/8	VH9071
40		3-7/8	VH9073
45	1-31/32 x 3-21/32	3-7/8	VH9115
50		3-7/8	VH9075

40 VAC, 50/60 Hz, Running Capacitors

Typical Ratings*	Dimensions**		Model Number***
	Oval	Length	
10	1-5/16 x 2-5/32	3-7/8	VH5300
15	1-29/32 x 2-29/32	2-7/8	VH5304
17.5	1-29/32 x 2-29/32	3-7/8	VH9141
20	1-29/32 x 2-29/32	3-7/8	VH9082
25	1-29/32 x 2-29/32	3-7/8	VH5310
30		4-3/4	VH9086
35	1-29/32 x 2-29/32	4-3/4	VH9088
40		4-3/4	VH9641
45		3-7/8	VH5351
50	1-31/32 x 3-21/32	3-7/8	VH5320
55		4-3/4	VH9081

* in μF
** in inches
*** Model numbers vary by manufacturer

STANDARD WIRING COLOR CODES

Electronic Applications (established by the Electronic Industries Association):

Insulation Color	Circuit type
Black	Chassis grounds, returns, primary leads
Blue	Plate leads, transistor collectors, FET drain
Brown	Filaments, plate start lead
Gray	AC main power leads
Green	Transistor base, finish grid, diodes, FET gate
Orange	Transistor base 2, screen grid
Red	B plus dc power supply
Violet	Power supply minus
White	B – C minus of bias supply, AVC – AGC return
Yellow	Emitters-cathode and transistor, FET source

Stereo Audio Channels:

Insulation Color	Circuit type
White	Left channel high side
Blue	Left channel low side
Red	Right channel high side
Green	Right channel low side

AF Transformers (Audio):

Insulation Color	Circuit type
Black	Ground line
Blue	Plate, collector, or drain lead; End of primary winding
Brown	Start primary loop; Opposite to blue lead
Green	High side, end secondary loop
Red	B plus, center tap push-pull loop
Yellow	Secondary center tap

IF Transformers (Intermediate Frequency):

Insulation Color	Circuit type
Blue	Primary high side of plate, collector, or drain lead
Green	Secondary high side for output
Red	Low side of primary returning B plus
Violet	Secondary outputs
White	Secondary low side

SINE WAVES

Frequency	Period	Peak-to-Peak Value
$f = \dfrac{1}{T}$ where f = frequency (in hertz) 1 = constant T = period (in seconds)	$T = \dfrac{1}{f}$ where T = period (in seconds) 1 = constant f = frequency (in hertz)	$V_{p\text{-}p} = 2 \times V_{max}$ where 2 = constant $V_{p\text{-}p}$ = peak-to-peak value V_{max} = peak value

Average Value	rms Value
$V_{avg} = V_{max} \times .637$ where V_{avg} = average value (in volts) V_{max} = peak value (in volts) .637 = constant	$V_{rms} = V_{max} \times .707$ where V_{rms} = rms value (in volts) V_{max} = peak value (in volts) .707 = constant

CONDUCTIVE LEAKAGE CURRENT

$$I_L = \frac{V_a}{R_1}$$

where

I_L = leakage current (in microamperes)

V_a = applied voltage (in volts)

R_1 = insulation resistance (in megohms)

PROGRAMMABLE CONTROLLER ERROR CODES

Error Code	Problem	Possible Cause	Corrective Action
01	Memory error occuring in run mode	Voltage surge, improper grounding, high noise interference on lines, inadequate power supply	Add surge suppression to incoming power lines and any inductive output lines. Check output voltage of power supply. Reload program with backup program and reboot system.
02	Processor does not meet required software level	Software is not compatible with the hardware system. Normally occurs when updated software is loaded into an older system	Consult the software specifications or manufacturer to determine the required hardware level. Upgrade system hardware to meet the software requirements.
03	Power failure of an expansion I/O module	Power was removed from module, inadequate power supply, or power has dipped below the minimum specification of the module	Measure voltage at the module power supply and correct any problems. Reboot the system to return to normal operation when message still appears after power is restored.
04	Program is trying to address an I/O module in an empty slot on rack	Program is set to the wrong rack and/or slot number	Set the program to a new address number or insert the correct I/O module in the slot number being addressed.
05	Module has been detected as being inserted under power	Power was not disabled before a module was inserted into a slot	Reboot the system to return to normal operation when message still appears after power is restored. Never insert a module while power is applied to the unit. Always turn power OFF before inserting or removing a module.

WAVELENGTHS – ELECTROMAGNETIC

DESCRIPTION (BAND)	FREQUENCY
Non-visible Light Spectrum (Range)	.0005 Angstrom Units to .39 Micrometers (μM)
Cosmic Rays Gamma Rays X Rays Ultraviolet Light	.0005 Angstrom Units .0000006 to .00001 (μM) .00001 to .032 (μM) .032 to .39 (μM)
Visible Light Spectrum	.39 to 76 Micrometers (μM)
Violet Light Blue Light Green Light Maximum Visibility Yellow Light Orange Light Red Light	4240 to 4000 Angstrom Units 4912 to 4240 Angstrom Units 5960 to 4912 Angstrom Units 5750 to 5560 Angstrom Units 5850 to 5750 Angstrom Units 6740 to 5850 Angstrom Units 7000 to 6470 Angstrom Units
Infrared Heat & Light	.76 to 30 Micrometers (μM)
Extra-High Frequencies such as Gov't Experiments and Radar	300 Gigahertz (GHZ) to 30,000 - 3,000 Megahertz (MHZ)
Super High Frequencies such as Radio Navigation and Amateur Radio	30,000 to 3,000 Megahertz (MHZ)
Ultra High Frequencies	3000 to 300 Megahertz (MHZ)
Radars	1600 to 1300 (MHZ)
Non-Government and Government Radio Navigation, Amateur Radio Cellular Telephone Channels 14-69 (TV) Class A Citizens .7 Meter Amateur Radio	3000 to 890 (MHZ) 890 to 806 (MHZ) 806 to 470 (MHZ) 563.2 to 462.55 (MHZ) 420 to 400 (MHZ)
Very High Frequencies	300 to 30 Magahertz (MHZ)
United States Military 1¼ Meter Amateur Radio CB Radio, Navigation (AIR) Amateur Radio, Government & Non-Government Channels 7 – 13 (TV) 2 Meter Amateur Radio	400 to 275 (MHZ) 225 to 220 (MHZ) 470 to 216 (MHZ) 216 to 174 (MHZ) 148 to 144 (MHZ)

WAVELENGTHS – ELECTROMAGNETIC (cont.)

DESCRIPTION (BAND)	FREQUENCY
United States Government	174 to 148 (MHZ)
Communication (Civil)	136 to 118 (MHZ)
Navigation (Areonautical)	118 to 108 (MHZ)
FM Broadcast	108 to 88 (MHZ)
Channels 5 – 6 (TV)	88 to 76 (MHZ)
U.S. Gov't (Aeronautical)	76 to 72 (MHZ)
Channels 2 – 4 (TV)	72 to 54 (MHZ)
6 Meter Amateur Radio	54 to 50 (MHZ)
Railroads, Fire and Police	50 to 30 (MHZ)
High Frequencies	30 to 3 Megahertz (MHZ)
Class D Citizens	27.405 to 26.965 (MHZ)
Science, Medicine and Industries	27.54 to 26.95 (MHZ)
10 Meter Amateur Radio	29.7 to 28 (MHZ)
12 Meter Amateur Radio	24.99 to 24.89 (MHZ)
15 Meter Amateur Radio	21.45 to 21 (MHZ)
17 Meter Amateur Radio	18.168 to 18.068 (MHZ)
20 Meter Amateur Radio	14.35 to 14.1 (MHZ)
30 Meter Amateur Radio	10.15 to 10.1 (MHZ)
40 Meter Amatuer Radio	7.3 to 7 (MHZ)
80 Meter Amateur Radio	4 to 3.5 (MHZ)
Medium Frequencies	3000 to 300 Kilohertz (KHZ)
AM Broadcast	1705 to 535 (KHZ)
160 Meter Amateur Radio	2000 to 1800 (KHZ)
Low Frequencies	300 to 30 Kilohertz (KHZ)
Communication/Navigation (Marine)	535 to 30 (KHZ)
All Radio Frequencies	30,000 MHZ to 30 (KHZ)
Very Low Frequency	30 KHZ to 3 (KHZ)
Ultrasonic Frequency	16 KHZ to 10 (KHZ)
Ultra Low Frequency	300 HZ to 30 HZ
Extremely Low Frequency	30 HZ to 3 HZ

WAVELENGTHS – MECHANICAL

DESCRIPTION (BAND)	FREQUENCY – HERTZ (HZ)
Normal Music	4186.01 to 16 HZ
Standard Range of Hearing (Human)	15000 to 30 HZ
Audio	20000 to 15 HZ
Direct Current (Continuous)	0 Hertz

SMALL TUBE FUSES

CHARACTERISTICS	Fuse Type	Fuse Diameter	Fuse Length
Dual Element, time delay, glass tube	MDL	1/4"	1-1/4"
Dual Element, glass tube	MDX	1/4"	1-1/4"
Dual Element, glass tube, pigtail	MDV	1/4"	1-1/4"
Ceramic Body, normal, 200% 15 sec	3AB	1/4"	1-1/4"
Metric, fast blow, high int., 210% 30 minutes	216	5mm	20mm
Glass, metric, fast blow, 210% 30 minutes	217	5mm	20mm
Glass, metric, slow blow, 210% 2 minutes	218	5mm	20mm
No Delay, ceramic, 110% rating, opens at 135% load in one hour	ABC	1/4"	1-1/4"
Fast Acting, glass tube, 110% rating, opens at 135% load in one hour	AGC	1/4"	1-1/4"
Fast Acting, glass tube	AGX	1/4"	1"
No Delay, 200% 15 sec	BLF	13/32"	1-1/2"
No Delay, military, 200% 15 sec	BLN	13/32"	1-1/2"
Fast Cleaning, 600V, 135% 1 hr	BLS	13/32"	1-3/8"
Time Delay, indicator pin, 135% 1 hr	FLA	13/32"	1-1/2"
Dual Element, delay, 200% 12 sec	FLM	13/32"	1-1/2"
Dual Element, delay, 500V, 200% 12 sec	FLQ	13/32"	1-1/2"
Slow Blow time delay	FNM	13/32"	1-1/4"
Slow Blow, indicator, metal pin pops outs indicating blown, dual element	FNA	13/32"	1-1/2"
Rectifier Fuse, fast, low let through	GBB	1/4"	1-1/4"
Indicator Fuse, metal pin pops out indicating blown, 110% rating	GLD	1/4"	1-1/4"
Metric, fast acting	GGS	5mm	20mm
Fast, current limiting, 600V, 135% 1 hr	KLK	13/32"	1-1/2"
Fast, protect solid state, 250% 1 sec	KLW	13/32"	1-1/2"
Slow Blow, time delay size rejection also	SC	13/32"	1-5/6" to 2-1/4"
Slow Blow, glass body, 200% 5 sec	218000	0.197"	0.787"
Slow Blow, glass body, 200% 5 sec	313000	1/4"	1-1/4"
Slow Blow, ceramic, 200% 5 sec	326000	1/4"	1-1/4"
Auto Glass, fast blow, 200% 5 sec	1AG	1/4"	5/8"
Auto Glass, fast blow, 200% 10 sec	2AG	0.177"	0.57"
Auto Glass, fast blow, 200% 5 sec	3AG	1/4"	1-1/4"
Auto Glass, fast blow, 200% 5 sec	4AG	9/32"	1-1/4"
Auto Glass, fast blow, 200% 5 sec	8AG	13/32"	1-1/2"
Auto Glass, fast blow, 200% 5 sec	7AG	1/4"	7/8"
Auto Glass, fast blow, 200% 5 sec	8AG	1/4"	1"
Auto Glass, fast blow, 200% 5 sec	9AG	1/4"	1-7/16"

The percentage and time figures mean that a 135% overload will blow a KLK type fuse in 1 hour. (example)

BASIC BATTERY SIZES

TYPE AND SIZE	Voltage	NEDA #	Capacity
Ni Cad Rechargeable:			
AAA	1.2	10024	180 milliamp-hours
AA	1.2	10015	500 milliamp-hours
C	1.2	10014	1.2 ampere-hours
Sub C	1.2	10022	1.2 ampere-hours
D	1.2	10013	1.2 ampere-hours
D	1.2	10013HC	4 ampere-hours
N	1.2	10910	150 milliamp-hours
Alkaline–Manganese:			
AAA	1.5	24A	37.5 ma @ 25 hours
AA	1.5	15A	20 ma @ 107 hours
C	1.5	14A	37.5 ma @ 160 hours
D	1.5	13A	50 ma @ 270 hours
G	6.0	930A	375 ma @ 59 hours
N	1.5	910A	9 ma @ 90 hours
	3.0	1308AP	20 ma @ 35 hours
	4.5	1307AP	20 ma @ 35 hours
	9.0	1604A	18 ma @ 33 hours
Carbon Zinc Cells:			
AAA	1.5	24F	20 ma @ 21 hours
AA	1.5	15F	54 ma @ 20 hours
C	1.5	14F	20 ma @ 140 hours
C	1.5	14D	37.5 ma @ 97 hours
D	1.5	13F	20 ma @ 360 hours
D	1.5	13C	375 ma @ 15.8 hours
D	1.5	13D	60 ma @ 139 hours
N	1.5	910F	20 ma @ 22 hours
WO	1.5	-	0.1 ma @ 650 hours
	3.0	704	20 ma @ 37 hours
	4.5	903	120 ma @ 90 hours
	6.0	2	60 ma @ 175 hours
	6.0	908	187 ma @ 40 hours
109	9	1611	12 ma @ 40 hours
127	9	1600	12 ma @ 61 hours
	9	1603	20 ma @ 350 hours
117	9	1604	9 ma @ 50 hours
	12	1810	12 ma @ 59 hours
	22.5	225	5 ma @ 60 hours
	45	207	40 ma @ 125 hours
	90	204	10 ma @ 63 hours

FUEL CELLS AND BATTERIES

TYPE OF UNIT Fuel Cells	CATHODE (−)	ANODE (+)	ACTUAL WORKING VOLTAGE (WV)	TOTAL VOLTAGE (TV)	AMPERE HOURS PER KG.
Hydrogen	O_2	H_2	0.7	1.23	26,000
Hydrazine	O_2	N_2H_4	0.7	1.5	2,100
Methanol	O_2	CH_2OH	0.9	1.3	1,400
Batteries					
Cadm-Air (RC)	O_2	Cd	0.8	1.2	475
Edison (RC)	NiO.	Fe	1.2	1.5	195
Hyd.Perox. (RC)	O_2	H_2	0.8	1.23	3,000
Lead-Acid (RC)	PbO_2	Pb	2.0	2.1	55
NiCad (RC)	NiO	Cd	1.2	1.35	165
Silver Cadm. (RC)	AgO	Cd	1.05	1.4	230
Ammonia	m-DNB	Mg	1.7	2.2	1,400
Cuprous Chloride	CuCl	Mg	1.4	1.5	240
Leclance	MnO_2	Zn	1.2	1.6	230
Lithium High-Temp with fused salt	S	Li	1.8	2.1	685
Magnesium	MnO_2	Mg	1.5	2.0	270
Mercury	HgO	Zn	1.2	1.34	185
Mercad	HgO	Cd	0.85	0.9	165
MnO_2	MnO_2	Zn	1.15	1.5	230
Organic Cath.	mDNB	Mg	1.15	1.8	1,400
Silver Chloride	AgCl	Mg	1.5	1.6	170
Silver Oxide	AgO	Zn	1.5	1.85	285
Silver-Poly	Polyiodide	Ag	0.6	0.66	180
Sodium High-Temp with electrolyte	S	Na	1.8	2.2	1,150
Thermal	Fuel	Ca	2.6	2.8	240
Zinc-Air	O_2	Zn	1.1	1.6	815
Zinc-Nickel	Ni oxides	Zn	1.6	1.75	185
Zinc-Silver Oxide	AgO	Zn	1.5	1.85	285

(RC) Indicates the cell is secondary and can be recharged. (WV) Indicates the average voltage generated by a working cell. (TV) Indicates the theoretical voltage developed by the cell. Ampere hours per KG. is the theoretical capacity of the cell.

2-14

CHARACTERISTICS OF LEAD – ACID BATTERIES

TEMPERATURE (F) VERSUS BATTERY EFFICIENCY (%)

−20° — 18%	20° — 58%
−10° — 33%	30° — 64%
0° — 40%	50° — 82%
10° — 50%	80° —100%

CHARGE	SPECIFIC GRAVITY OF ACID
Discharged	1.11 to 1.12
Very Low Capacity	1.13 to 1.15
25% of Capacity	1.15 to 1.17
50% of Capacity	1.20 to 1.22
75% of Capacity	1.24 to 1.26
100% of Capacity	1.26 to 1.28
Overcharged	1.30 to 1.32

MAGNETIC PERMEABILITY OF SOME COMMON MATERIALS

Substance	Permeability (approx.)
Aluminum	Slightly more than 1
Bismuth	Slightly less than 1
Cobalt	60-70
Ferrite	100-300
Free space	1
Iron	60-100
Iron, refined	3000-8000
Nickel	50-60
Permalloy	3000-30,000
Silver	Slightly less than 1
Steel	300-600
Super permalloys	100,000-1,000,000
Wax	Slightly less than 1
Wood, dry	Slightly less than 1

TRANSISTOR CIRCUIT ABBREVIATIONS

Quantity	Abbreviations
Base-emitter voltage	E_B, V_B, E_{BE}, V_{BE}
Collector-emitter voltage	E_C, V_C, E_{CE}, V_{CE}
Collector-base voltage	$E_{BC}, V_{BC}, E_{CB} V_{CB}$
Gate-source voltage	E_G, V_G, E_{GS}, V_{GS}
Drain-source voltage	E_D, V_D, E_{DS}, V_{DS}
Drain-gate voltage	$E_{DG}, V_{DG}, E_{DG}, V_{DG}$
Emitter current	I_E
Base current	I_B, I_{BE}, I_{EB}
Collector current	I_C, I_{CE}, I_{EC}
Source current	I_S
Gate current	I_G, I_{GS}, I_{SG}^*
Drain current	I_D, I_{DS}, I_{SD}

*This is almost always significant.

RADIO FREQUENCY CLASSIFICATIONS

Classification	Abbreviation	Frequency range
Very Low Frequency	VLF	9 kHz and below
Low Frequency (Longwave)	LF	30 kHz - 300 kHz
Medium Frequency	MF	300 kHz - 2MHz
High Frequency (Shortwave)	HF	3 MHz - 30MHz
Very High Frequency	VHF	30 MHz - 300 MHz
Ultra High Frequency	UHF	300 MHz - 3 GHz
Microwaves	MW	3 GHz and more

ELECTRONIC SYMBOLS

ammeter	
amplifier (operational)	
AND gate	
antenna (balanced, dipole)	
antenna (general)	
antenna (shielded)	
antenna (unshielded)	
antenna (unbalanced)	
antenna (whip)	
attenuator (or resistor, fixed)	
attenuator (or resistor, variable)	
battery	

capacitor (feedthrough)	⊥ ⊣⏐⊢
capacitor (fixed, nonpolarized)	⊁⏐
capacitor (fixed, polarized)	⊣⏐⊢
capacitor (ganged, variable)	⊁--⊁
capacitor (single variable)	⊁
capacitor (split-rotor, variable)	⊁
capacitor (split-stator, variable)	⊣⏐⏐⊢
cathode (directly heated)	∩
cathode (indirectly heated)	⊓ ⌐
cathode (cold)	Y
cavity resonator	⊟
cell	⊣⏐⊢

ELECTRONIC SYMBOLS (cont.)

coaxial cable	
coaxial cable (grounded shield)	
crystal (piezoelectric)	
delay line	
diode (field effect)	
diode (general)	
diode (Gunn)	
diode (light-emitting)	
diode (photosensitive)	
diode (photovoltaic)	
diode (pin)	
diode (Schottky)	

diode (tunnel)

diode (varactor)

diode (zener)

directional coupler (or wattmeter)

exclusive-OR gate

female contact (general)

ferrite bead

fuse

galvanometer

ground (chassis)

ground (earth)

handset

headphone (single)	
headphone (stereo)	
inductor (air-core)	
inductor (bifilar)	
inductor (iron-core)	
inductor (tapped)	
inductor (variable)	
integrated circuit	
inverter or inverting amplifier	
jack (coaxial or phono)	
jack (phone, two-conductor)	
jack (phone, two-conductor interrupting)	

jack (phone, three-conductor)

jack (phono)

key (telegraph)

lamp (incandescent)

lamp (neon)

male contact (general)

meter (general)

microammeter

microphone

microphone (directional)

milliammeter

NAND gate

negative voltage connection

NOR gate

operational amplifier

OR gate

outlet (nonpolarized)

outlet (polarized)

outlet (utility, 117 V, nonpolarized)

outlet (utility, 234 V)

photocell (tube)

plug (nonpolarized)

plug (polarized)

plug (phone, two-conductor)

plug (phone, three-conductor)	
plug (phono)	
plug (utility, 117 V)	
plug (utility, 234V)	
positive-voltage connection	
potentiometer (variable resistor, or rheostat)	
probe (radio-frequency)	
rectifier (semiconductor)	
rectifier (silicon-controlled)	
rectifier (tube-type)	
rectifier (tube-type, gas-filled)	
relay (DPDT)	

ELECTRONIC SYMBOLS (cont.)

relay (DPST)

relay (SPDT)

relay (SPST)

resistor (fixed)

resistor (preset)

resistor (tapped)

resonator

rheostat (variable resistor, or potentiometer)

saturable reactor

shielding

signal generator

solar cell source (constant-voltage)

ELECTRONIC SYMBOLS (cont.)

source (constant-current)

speaker

switch (DPDT)

switch (DPST)

switch (momentary-contact)

switch (rotary)

switch (silicon-controlled)

switch (SPDT)

switch (SPST)

terminals (general, balanced)

terminals (general, unbalanced)

test point

ELECTRONIC SYMBOLS (cont.)

thermocouple	
thyristor (diac)	
thyristor (triac)	
transformer (air-core)	
transformer (air-core, adjustable)	
transformer (iron-core)	
transformer (iron-core, adjustable)	
transformer (powdered iron-core)	
transformer (tapped-primary)	
transformer (tapped-secondary)	
transistor (bipolar, npn)	
transistor (bipolar, pnp)	

ELECTRONIC SYMBOLS (cont.)

transistor (junction field-effector, JFET)

transistor (field-effect, n-channel)

transistor (field-effect, P-channel)

transistor (metal-oxide, dual-gate)

transistor (metal-oxide, single-gate)

transistor (photosensitive)

transistor (unijunction)

tube (diode)

tube (pentode)

tube (photomultiplier)

tube (tetrode)

tube (triode)

ELECTRONIC SYMBOLS (cont.)

unspecified unit or component

voltmeter

wattmeter

waveguide (circular)

waveguide (rectangular)

waveguide (flexible)

waveguide (twisted)

CHAPTER 3
RACEWAYS AND WIRING

TYPICAL POWER WIRING COLOR CODE

120/240 Volt		277/480 Volt	
Black	Phase 1	Brown	Phase 1
Red	Phase 2	Orange	Phase 2
Blue	Phase 3	Yellow	Phase 3
White or with 3 White stripes	Neutral	Gray or with 3 White stripes	Neutral
Green	Ground	Green w/Yellow stripe	Ground

POWER TRANSFORMERS ARE COLOR CODED AS FOLLOWS:

Wire Color (solid)	Circuit Type
Black	If a transformer does not have a tapped primary, both leads are black.
Black	If a transformer does have a tapped primary, the black is the common lead.
Black & Yellow .	Tap for a tapped primary.
Black & Red...	End for a tapped primary.

RESISTIVITIES OF DIFFERENT SOILS

Soil	Resistivity OHM-CM		
	Average	Min.	Max.
Fills – ashes, cinders, brine wastes	2,370	590	7,000
Clay, shale, gumbo, loam	4,060	340	16,300
Same – with varying proportions of sand and gravel	15,800	1,020	135,000
Gravel, sand, stones, with little clay or loam	94,000	59,000	458,000

BENDING STUB-UPS

Conduit or EMT Size	Deduction from Mark or Arrow
1/2"	5"
3/4"	6"
1"	8"

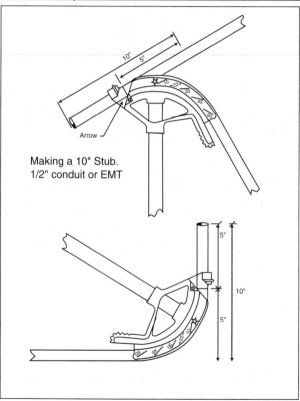

Making a 10" Stub.
1/2" conduit or EMT

BACK TO BACK BENDING

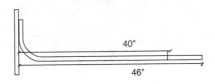

40"

46"

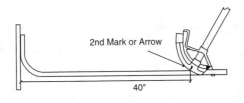

2nd Mark or Arrow

40"

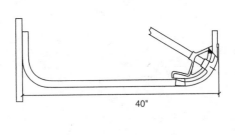

40"

OFFSET AND SADDLE BENDS

2-Bend Offset

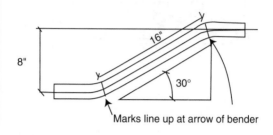

Marks line up at arrow of bender

Angle of bends	Inches of run per inch of offset	Loss of conduit length per inch of offset
10°	5.76	1/16"
22.5°	2.6	3/16"
30°	2.0	1/4"
45°	1.414	3/8"
60°	1.15	1/2"

3-Bend Saddle

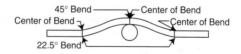

AMPACITY OF LAMP AND EXTENSION CORDS (AWG) - TYPES S, SJ, SJT, SP, SPT, ST

Gauge	4 Conductors	3 Conductors	2 Conductors
18	6	7	10
16	8	10	13
14	12	15	18
12	16	20	25
10	20	25	30

ENCLOSURES

Type	Service Conditions	Sealing Method	Cost
1	No unusual		Base
4	Windblown dust and rain, splashing water, hose-directed water, and ice on enclosure		12 x Base
4X	Corrosion, windblown dust and rain, splashing water, hose-directed water, and ice on enclosure		12 x Base
7	Withstand and contain an internal explosion of specified gases, contain an explosion sufficiently so an explosive gas-air mixture in the atmosphere is not ignited		48 x Base
9	Dust		48 x Base
12	Dust, falling dirt, and dripping noncorrosive liquids		5 x Base

HAZARDOUS LOCATIONS

Class	Group	Material
I	A	Acetylene
	B	Hydrogen, butadiene, ethylene oxide, propylene oxide
	C	Carbon monoxide, ether, ethylene, hydrogen sulfide, morpholine, cyclopropane
	D	Gasoline, benzene, butane, propane, alcohol, acetone, ammonia, vinyl chloride
II	E	Metal dusts
	F	Carbon black, coke dust, coal
	G	Grain dust, flour, starch, sugar, plastics
III	No groups	Wood chips, cotton, flax, and nylon

ENCLOSURE TYPES

Type	Use	Service Conditions	UL Tests	Comments
1	Indoor	None	Rod entry, rust resistance	
3	Outdoor	Windblown dust, rain, sleet, and ice on enclosure	Rain, external icing, dust, and rust resistance	Do not provide protection against internal condensation, or internal icing
3R	Outdoor	Falling rain and ice on enclosure	Rod entry, rain, external icing, and rust resistance	Do not provide protection against dust, internal condensation or internal icing
4	Indoor/outdoor	Windblown dust and rain, splashing water, hose-directed water and ice on enclosure	Hosedown, external icing and rust resistance	Do not provide protection against internal condensation or internal icing
4X	Indoor/outdoor	Corrosion, windblown dust and rain, splashing water, hose-directed water and ice on enclosure	Hosedown, external icing, and corrosion resistance	Do not provide protection against internal condensation or internal icing
6	Indoor/outdoor	Occasional temporary submersion at a limited depth		
6P	Indoor/outdoor	Prolonged submersion at a limited depth		

ENCLOSURE TYPES *(cont.)*

Type	Use	Service Conditions	UL Tests	Comments
7	Indoor locations classified as Class I, or Groups A, B, C, or D, as defined in the NEC®	Withstand and contain an internal explosion of specified gases, contain an explosion sufficiently so an explosive gas-air mixture in the atmosphere is not ignited	Explosion, hydrostatic, and temperature	Enclosed heat-generating devices shall not cause external surfaces to reach temperatures capable of igniting explosive gas-air mixtures in the atmosphere
9	Indoor locations classified as Class II, Groups E or G, as defined in the NEC®	Dust	Dust penetration, temperature, and gasket aging	Enclosed heat-generating devices shall not cause external surfaces to reach temperatures capable of igniting explosive gas-air mixtures in the atmosphere
12	Indoor	Dust, falling dirt, and dripping noncorrosive liquids	Drip, dust, and rust resistance	Do not provide protection against internal condensation
13	Indoor	Dust, spraying water, oil and noncorrosive coolant	Oil explosion and rust resistance	Do not provide protection against internal condensation

3-7

SIZES OF PANELBOARDS

SINGLE-PHASE — 3-WIRE SYSTEMS

40 A	100 A	150 A	225 A	400 A
70 A	125 A	200 A	300 A	600 A

THREE-PHASE — 4-WIRE SYSTEMS

60 A	150 A	225 A	400 A
125 A	200 A	300 A	600 A

SIZES OF GUTTERS and WIREWAYS

2-1/2" x 2-1/2" 4" x 4"	6" x 6" 8" x 8"	10" x 10"

These sizes are available in 12", 24", 36", 48", and 60" lengths.

SIZES OF DISCONNECTS

30 A	200 A	800 A	1600 A
60 A	400 A	1200 A	1800 A
100 A	600 A	1400 A	

SIZES OF PULL BOXES and JUNCTION BOXES

4" x 4" x 4"	10" x 8" x 4"	12" x 12" x 6"
6" x 4" x 4"	10" x 8" x 6"	12" x 12" x 8"
6" x 6" x 4"	10" x 10" x 4"	15" x 12" x 4"
6" x 6" x 6"	10" x 10" x 6"	15" x 12" x 6"
8" x 6" x 4"	10" x 10" x 8"	18" x 12" x 4"
8" x 6" x 6"	12" x 8" x 4"	18" x 12" x 6"
8" x 6" x 8"	12" x 8" x 6"	18" x 18" x 4"
8" x 8" x 4"	12" x 10" x 4"	18" x 18" x 6"
8" x 8" x 6"	12" x 10" x 6"	24" x 18" x 6"
8" x 8" x 8"	12" x 12" x 4"	24" x 24" x 6"
		24" x 24" x 8"

BUSWAY or BUSDUCT	
1 φ	3 φ
225 A	225 A
400 A	400 A
600 A	600 A
800 A	800 A
1000 A	1000 A
1200 A	1200 A
1350 A	1350 A
1600 A	1600 A
2000 A	2000 A
2500 A	2500 A
3000 A	3000 A
4000 A	4000 A
5000 A	5000 A

CBs and FUSES			
15	70	225	800
20	80	250	1000
25	90	300	1200
30	100	350	1600
35	110	400	2000
40	125	450	2500
45	150	500	3000
50	175	600	4000
60	200	700	5000
			6000

For fuses only, additional standard sizes are 1, 3, 6, and 10.
All sizes are rated in amps.

SWITCHBOARDS or SWITCHGEARS	
1 φ	3 φ
200 A	400 A
400 A	600 A
600 A	800 A
800 A	1200 A
1200 A	1600 A
1600 A	2000 A
2000 A	2500 A
2500 A	3000 A
3000 A	4000 A
4000 A	

ELECTRICAL CABLE CLASS RATINGS

Electrical cable is rated according to the following parameters. The number of wires, the wire size, the type of insulation and the moisture condition of the environment of the wire. Therefore, an electrical cable designated 10/2 with ground - type UF - 600volt-(UL) meets the specifications below:

- The "10" relates to wire size - 10 gauge wire.
- 2 defines an electrical cable with two wires.
- The word "Ground" indicates the cable has a third wire to be connected to ground.
- The term "Type UF" means the insulation type has an acceptable moisture rating.
- "600 V" defines the cable as being rated at 600 volts maximum.
- ("UL") means the cable has certification from Underwriters Laboratory.

CABLE INSULATION MOISTURE RATINGS

DRY – Indoor above ground level; moisture usually not encountered.

MOIST – Indoor below ground level (basement); locations are partially protected; moisture level is moderate.

WET – Locations affected by weather (outside); concrete slabs, underground, etc.; water saturation likely.

CONDUCTOR PREFIX CODES

B	– Outer braid	O	– Neoprene jacket
F	– Fixture wire	R	– Rubber covering
FEP	– Fluorinated ethylene propylene. Use in dry locations only, over 90° C	S	– Appliance cord
		SP	– Lamp cord, rubber
H	– Load temp up to 75° C	SPT	– Lamp cord, plastic
HH	– Load temp up to 90° C	T	– Load temp up to 60° C
L	– Seamless lead jacket	W	– Wet use only
M	– Machine tool wire	X	– Moisture and heat resistant
N	– Resistant to oil and gas		

TYPES OF CONDUCTORS

TYPE	MAX. TEMP	APPLICATION	INSULATION	OUTER COVERING
FEP or FEPB	90°C (194°F) 200°C (392°F)	Dry and damp locations Dry locations – Special Apps	Fluorinated ethylene propylene	None or glass braid
MI	90°C (194°F) 250°C (482°F)	Dry and wet locations Special Apps	Magnesium oxide	Copper or alloy steel
MTW	60°C (140°F) 90°C (194°F)	Machine tool wiring – wet locations Machine tool wiring – dry locations	Flame-retardant, moisture, heat, and oil-resistant thermoplastic	None or nylon jacket
PAPER	85°C (185°F)	Underground service conductors	Paper	Lead sheath
PFA	90°C (194°F) 200°C (392°F)	Dry and damp locations Dry locations-Special Apps	Perfluoroalkoxy	None
PFAH	250°C (482°F)	Dry locations only	Perfluoroalkoxy	None
RHH	90°C (194°F)	Dry and damp locations		Moisture resistant, flame-retardant non-metallic
RHW	75°C (167°F)	Dry and wet locations	Flame-retardant, moisture- resistant thermoset	Moisture-resistant, flame-retardant, non-metallic

3-11

TYPES OF CONDUCTORS

TYPE	MAX. TEMP	APPLICATION	INSULATION	OUTER COVERING
RHW-2	90°C (194°F)	Dry and wet locations	Flame-retardant, moisture-resistant thermoset	Moisture-resistant, flame-retardant, non-metallic
SA	90°C (194°F) 200°C (392°F)	Dry and damp locations Special Apps	Silicone rubber	Glass or braid material
SIS	90°C (194°F)	Switchboard wiring	Flame-retardant thermostat	None
TBS	90°C (194°F)	Switchboard wiring	Thermoplastic	Flame-retardant, non-metallic
TFE	250°C (482°F)	Dry locations only	Extruded polytetrafluoroethylene	None
THHN	90°C (194°F)	Dry and damp locations	Flame-retardant, heat-resistant thermoplastic	Nylon jacket
THHW	75°C (167°F) 90°C (194°F)	Wet locations Dry locations	Flame-retardant, moisture- and heat-resistant thermoplastic	None
THW	75°C (167°F) 90°C (194°F)	Dry and wet locations Special Apps	Flame-retardant, moisture- and heat-resistant thermoplastic	None
THWN	75°C (167°F)	Dry and wet locations	Flame-retardant, moisture- and heat-resistant thermoplastic	Nylon jacket
TW	60°C (140°F)	Dry and wet locations	Flame-retardant, moisture- and heat-resistant thermoplastic	None
UF	60°C (140°F) 75°C (167°F)	Refer to NEC®	Moisture-resistant Moisture- and heat-resistant	Integral with insulation

TYPES OF CONDUCTORS

TYPE	MAX. TEMP	APPLICATION	INSULATION	OUTER COVERING
USE	75°C (167°F)	Refer to NEC®	Heat and moisture-resistant	Moisture-resistant non-metallic
XHH	90°C (194°F)	Dry and damp locations	Flame-retardant thermoplastic	None
XHHW	90°C (194°F) 75°C (167°F)	Dry and damp locations Wet locations	Flame-retardant, moisture-resistant thermoset	None
XHHW-2	90°C (194°F)	Dry and wet locations	Flame-retardant, moisture-resistant thermoset	None
Z	90°C (194°F) 150°C (302°F)	Dry and damp locations Dry locations – Special Apps	Modified ethylene tetrafluoro ethylene	None
ZW	75°C (167°F) 90°C (194°F) 150°C (302°F)	Wet locations Dry and damp locations Dry locations – Special Apps	Modified ethylene tetrafluoro ethylene	None

JUNCTION BOX CALCULATIONS

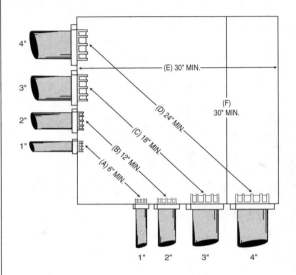

Distance (A) is 6 x 1" = 6" minimum
Distance (B) is 6 x 2" = 12" minimum
Distance (C) is 6 x 3" = 18" minimum
Distance (D) is 6 x 4" = 24" minimum
Distance (E) is (6 x 4") + 3" + 2" + 1" = 30" minimum
Distance (F) is (6 x 4") + 3" + 2" + 1" = 30" minimum

BOX FILL

Max. Number of Conductors in Outlet, Device and Junction Boxes

Box Dimension in Inches Trade Size or Type	Min. Capacity (in.³)	Maximum Number of Conductors						
		No. 18	No. 16	No. 14	No. 12	No. 10	No. 8	No. 6
4 x 1-1/4 round or octagonal	12.5	8	7	6	5	5	5	2
4 x 1-1/2 round or octagonal	15.5	10	8	7	6	6	5	3
4 x 2-1/8 round or octagonal	21.5	14	12	10	9	8	7	4
4 x 1-1/4 square	18.0	12	10	9	8	7	6	3
4 x 1-1/2 square	21.0	14	12	10	9	8	7	4
4 x 2-1/8 square	30.3	20	17	15	13	12	10	6
4-11/16 x 1-1/4 square	25.5	17	14	12	11	10	8	5
4-11/16 x 1-1/2 square	29.5	19	16	14	13	11	9	5
4-11/16 x 2-1/8 square	42.0	28	24	21	18	16	14	8
3 x 2 x 1-1/2 device	7.5	5	4	3	3	3	2	1
3 x 2 x 2 device	10.0	6	5	5	4	4	3	2
3 x 2 x 2-1/4 device	10.5	7	6	5	4	4	3	2
3 x 2 x 2-1/2 device	12.5	8	7	6	5	5	4	2
3 x 2 x 2-3/4 device	14.0	9	8	7	6	5	4	2
3 x 2 x 3-1/2 device	18.0	12	10	9	8	7	6	3
4 x 2-1/8 x 1-1/2 device	10.3	6	5	5	4	4	3	2
4 x 2-1/8 x 1-7/8 device	13.0	8	7	6	5	5	4	2
4 x 2-1/8 x 2-1/8 device	14.5	9	8	7	6	5	4	2
3-3/4 x 2 x 2-1/2 masonry box/gang	14.0	9	8	7	6	5	4	2
3-3/4 x 2 x 3-1/2 masonry box/gang	21.0	14	12	10	9	8	7	2

If one or more cable clamp is in a box, it is counted the same as the largest conductor. A loop of conductor 12 inches or more counts as two conductors.

Conductor Volume Allowance

Wire Size (AWG)	Volume Each (In.³)	Formula
18	1.50	$V = L \times W \times D$
16	1.75	
14	2.00	Volume =
12	2.25	Length times width
10	2.50	times depth
8	3.00	(in cubic inches)
6	5.00	

To find box size needed, add up total volume for all wires to be used. Then use the volume formula. Example: If total volume of all wires is 420 cubic inches — use an 8" x 10" x 6" box = 480 cubic inches.

INSULATED CONDUCTOR DIMENSIONS IN SQUARE INCHES

TYPE	SIZE	IN.^{2}	TYPE	SIZE	IN.^{2}
RFH-2,	18	0.0145	RHH*, RHW*, RHW-2*	14	0.0209
FFH-2	16	0.0172		12	0.0260
RHW-2, RHH,	14	0.0293	THHW, THW, AF,	10	0.0333
RHW	12	0.0353	XF, XFF		
	10	0.0437	RHH*, RHW*, RHW-2*	8	0.0556
	8	0.0835			
	6	0.1041	TW, THW,	6	0.0726
	4	0.1333	THHW,	4	0.0973
	3	0.1521	THW-2,	3	0.1134
	2	0.1750	RHH*,	2	0.1333
	1	0.2660	RHW*,	1	0.1901
	1/0	0.3039	RHW-2*	1/0	0.2223
	2/0	0.3505		2/0	0.2624
	3/0	0.4072		3/0	0.3117
	4/0	0.4754		4/0	0.3718
	250	0.6291		250	0.4596
	300	0.7088		300	0.5281
	350	0.7870		350	0.5958
	400	0.8626		400	0.6619
	500	1.0082		500	0.7901
	600	1.2135		600	0.9729
	700	1.3561		700	1.1010
	750	1.4272		750	1.1652
	800	1.4957		800	1.2272
	900	1.6377		900	1.3561
	1000	1.7719		1000	1.4784
	1250	2.3479		1250	1.8602
	1500	2.6938		1500	2.1695
	1750	3.0357		1750	2.4773
	2000	3.3719		2000	2.7818
SF-2, SFF-2	18	0.0115	TFN,	18	0.0055
	16	0.0139	TFFN	16	0.0072
	14	0.0172	THHN,	14	0.0097
SF-1, SFF-1	18	0.0065	THWN,	12	0.0133
RFH-1, XF, XFF	18	0.0080	THWN-2	10	0.0211
TF, TFF, XF, XFF	16	0.0109		8	0.0366
TW, XF, XFF, THHW,	14	0.0139		6	0.0507
THW, THW-2				4	0.0824
TW, THHW,	12	0.0181		3	0.0973
THW, THW-2	10	0.0243		2	0.1158
	8	0.0437		1	0.1562

*Denotes types minus outer covering

INSULATED CONDUCTOR DIMENSIONS IN SQUARE INCHES

TYPE	SIZE	IN.²	TYPE	SIZE	IN.²
THHN,	1/0	0.1855	XHHW, ZW,	14	0.0139
THWN,	2/0	0.2223	XHHW-2,	12	0.0181
THWN-2	3/0	0.2679	XHH	10	0.0243
(cont.)	4/0	0.3237		8	0.0437
	250	0.3970		6	0.0590
	300	0.4608		4	0.0814
	350	0.5242		3	0.0962
	400	0.5863		2	0.1146
	500	0.7073	XHHW,	1	0.1534
	600	0.8676	XHHW-2,	1/0	0.1825
	700	0.9887	XHH	2/0	0.2190
	750	1.0496		3/0	0.2642
	800	1.1085		4/0	0.3197
	900	1.2311		250	0.3904
	1000	1.3478		300	0.4536
PF, PGFF, PGF, PFF,	18	0.0058		350	0.5166
PTF, PAF, PTFF, PAFF	16	0.0075		400	0.5782
PF, PGFF, PGF, PFF,	14	0.0100		500	0.6984
PTF, PAF, PTFF, PAFF,				600	0.8709
TFE, FEP, PFA,				700	0.9923
FEPB, PFAH				750	1.0532
TFE, FEP,	12	0.0137		800	1.1122
PFA, FEPB,	10	0.0191		900	1.2351
PFAH	8	0.0333		1000	1.3519
	6	0.0468		1250	1.7180
	4	0.0670		1500	2.0157
	3	0.0804		1750	2.3127
	2	0.0973		2000	2.6073
TFE, PFAH	1	0.1399	KF-2,	18	0.0031
TFE,	1/0	0.1676	KFF-2	16	0.0044
PFA,	2/0	0.2027		14	0.0064
PFAH, Z	3/0	0.2463		12	0.0093
	4/0	0.3000		10	0.0139
ZF, ZFF	18	0.0045	KF-1,	18	0.0026
	16	0.0061	KFF-1	16	0.0037
Z, ZF, ZFF	14	0.0083		14	0.0055
Z	12	0.0117		12	0.0083
	10	0.0191		10	0.0127
	8	0.0302			
	6	0.0430			
	4	0.0625			
	3	0.0855			
	2	0.1029			
	1	0.1269			

RIGID METALLIC CONDUIT-MAXIMUM NUMBER OF CONDUCTORS

TYPE	SIZE	RMC Trade Size (Inches)											
		1/2	3/4	1	1 1/4	1 1/2	2	2 1/2	3	3 1/2	4	5	6
RHH, RHW, RHW-2	14	4	7	12	21	28	46	66	102	136	176	276	398
	12	3	6	10	17	23	38	55	85	113	146	229	330
	10	3	5	8	14	19	31	44	68	91	118	185	267
	8	1	2	4	7	10	16	23	36	48	61	97	139
	6	1	1	3	6	8	13	18	29	38	49	77	112
	4	1	1	2	4	6	10	14	22	30	38	60	87
	3	1	1	1	4	5	9	12	19	26	34	53	76
	2	1	1	1	3	4	7	11	17	23	29	46	66
	1	0	1	1	1	3	5	7	11	15	19	30	44
	1/0	0	1	1	1	2	4	6	10	13	17	26	38
	2/0	0	1	1	1	2	4	5	8	11	14	23	33
	3/0	0	0	1	1	1	3	4	7	10	12	20	28
	4/0	0	0	1	1	1	3	4	6	8	11	17	24
	250	0	0	0	1	1	1	3	4	6	8	13	18
	300	0	0	0	1	1	1	2	4	5	7	11	16
	350	0	0	0	1	1	1	2	4	5	6	10	15
	400	0	0	0	1	1	1	1	3	4	6	9	13
	500	0	0	0	1	1	1	1	3	4	5	8	11
	600	0	0	0	0	1	1	1	2	3	4	6	9
	700	0	0	0	0	1	1	1	1	3	4	6	8
	750	0	0	0	0	0	1	1	1	3	3	5	8
	800	0	0	0	0	0	1	1	1	2	3	5	7
	1000	0	0	0	0	0	1	1	1	2	3	4	6
TW, THHW, THW, THW-2	14	9	15	25	44	59	98	140	216	288	370	581	839
	12	7	12	19	33	45	75	107	165	221	284	446	644
	10	5	9	14	25	34	56	80	123	164	212	332	480
	8	3	5	8	14	19	31	44	68	91	118	185	267
RHH*, RHW*, RHW-2*	14	6	10	17	29	39	65	93	143	191	246	387	558
	12	5	8	13	23	32	52	75	115	154	198	311	448
	10	3	6	10	18	25	41	58	90	120	154	242	350
	8	1	4	6	11	15	24	35	54	72	92	145	209
RHH*, RHW*, RHW-2*, TW, THHW, THW-2	6	1	3	5	8	11	18	27	41	55	71	111	160
	4	1	1	3	6	8	14	20	31	41	53	83	120
	3	1	1	3	5	7	12	17	26	35	45	71	103
	2	1	1	2	4	6	10	14	22	30	38	60	87
	1	1	1	1	3	4	7	10	15	21	27	42	61
	1/0	0	1	1	2	3	6	8	13	18	23	36	52
	2/0	0	1	1	2	3	5	7	11	15	19	31	44
	3/0	0	1	1	1	2	4	6	9	13	16	26	37
	4/0	0	0	1	1	1	3	5	8	10	14	21	31
	250	0	0	1	1	1	3	4	6	8	11	17	25
	300	0	0	1	1	1	2	3	5	7	9	15	22
	350	0	0	1	1	1	1	3	5	6	8	13	19
	400	0	0	0	1	1	1	3	4	6	7	12	17
	500	0	0	0	1	1	1	2	3	5	6	10	14
	600	0	0	0	1	1	1	1	3	4	5	8	12
	700	0	0	0	0	1	1	1	2	3	4	7	10
	750	0	0	0	0	1	1	1	2	3	4	7	10
	800	0	0	0	0	1	1	1	2	3	4	6	9
	1000	0	0	0	0	0	1	1	1	2	3	5	8
THHN, THWN, THWN-2	14	13	22	36	63	85	140	200	309	412	531	833	1202
	12	9	16	26	46	62	102	146	225	301	387	608	877
	10	6	10	17	29	39	64	92	142	189	244	383	552
	8	3	6	9	16	22	37	53	82	109	140	221	318
	6	2	4	7	12	16	27	38	59	79	101	159	230
	4	1	2	4	7	10	16	23	36	48	62	98	141
	3	1	1	3	6	8	14	20	31	41	53	83	120
	2	1	1	3	5	7	11	17	26	34	44	70	100
	1	1	1	1	4	5	8	12	19	25	33	51	74

NOTE:
(*) Denotes Types Minus Outer Covering

RIGID METALLIC CONDUIT-MAXIMUM NUMBER OF CONDUCTORS

TYPE	SIZE	RMC Trade Size (Inches)											
		1/2	3/4	1	1 1/4	1 1/2	2	2 1/2	3	3 1/2	4	5	6
THHN,	1/0	1	1	1	3	4	7	10	16	21	27	43	63
THWN,	2/0	0	1	1	2	3	6	8	13	18	23	36	52
THWN-2	3/0	0	1	1	1	3	5	7	11	15	19	30	43
(cont.)	4/0	0	1	1	1	2	4	6	9	12	16	25	36
	250	0	0	1	1	1	3	5	7	10	13	20	29
	300	0	0	1	1	1	3	4	6	8	11	17	25
	350	0	0	1	1	1	2	3	5	7	10	15	22
	400	0	0	1	1	1	2	3	5	7	8	13	20
	500	0	0	0	1	1	1	2	4	5	7	11	16
	600	0	0	0	1	1	1	1	3	4	6	9	13
	700	0	0	0	1	1	1	1	3	4	5	8	11
	750	0	0	0	1	1	1	1	3	4	5	7	11
	800	0	0	0	0	1	1	1	2	3	4	7	10
	1000	0	0	0	0	1	1	1	1	3	4	6	8
FEP, FEPB,	14	12	22	35	61	83	136	194	300	400	515	808	1166
PFA, PFAH,	12	9	16	26	44	60	99	142	219	292	376	590	851
TFE	10	6	11	18	32	43	71	102	157	209	269	423	610
	8	3	6	10	18	25	41	58	90	120	154	242	350
	6	2	4	7	13	17	29	41	64	85	110	172	249
	4	1	3	5	9	12	20	29	44	59	77	120	174
	3	1	2	4	7	10	17	24	37	50	64	100	145
	2	1	1	3	6	8	14	20	31	41	53	83	120
PFA, PFAH, TFE	1	1	1	2	4	6	9	14	21	28	37	57	83
PFA, PFAH,	1/0	1	1	1	3	5	8	11	18	24	30	48	69
TFE, Z	2/0	1	1	1	3	4	6	9	14	19	25	40	57
	3/0	0	1	1	2	3	5	8	12	16	21	33	47
	4/0	0	1	1	1	2	4	6	10	13	17	27	39
Z	14	15	26	42	73	100	164	234	361	482	621	974	1405
	12	10	18	30	52	71	116	166	256	342	440	691	997
	10	6	11	18	32	43	71	102	157	209	269	423	610
	8	4	7	11	20	27	45	64	99	132	170	267	386
	6	3	5	8	14	19	31	45	69	93	120	188	271
	4	1	3	5	9	13	22	31	48	64	82	129	186
	3	1	2	4	7	9	16	22	35	47	60	94	136
	2	1	1	3	6	8	13	19	29	39	50	78	113
	1	1	1	2	5	6	10	15	23	31	40	63	92
XHH,	14	9	15	25	44	59	98	140	216	288	370	581	839
XHHW,	12	7	12	19	33	45	75	107	165	221	284	446	644
XHHW-2,	10	5	9	14	25	34	56	80	123	164	212	332	480
ZW	8	3	5	8	14	19	31	44	68	91	118	185	267
	6	1	3	6	10	14	23	33	51	68	87	137	197
	4	1	2	4	7	10	16	24	37	49	63	99	143
	3	1	1	3	6	8	14	20	31	41	53	84	121
	2	1	1	3	5	7	12	17	26	35	45	70	101
XHH,	1	1	1	1	4	5	9	12	19	26	33	52	76
XHHW,	1/0	1	1	1	3	4	7	10	16	22	28	44	64
XHHW-2	2/0	0	1	1	2	3	6	9	13	18	23	37	53
	3/0	0	1	1	1	3	5	7	11	15	19	30	44
	4/0	0	1	1	1	2	4	6	9	12	16	25	36
	250	0	0	1	1	1	3	5	7	10	13	20	30
	300	0	0	1	1	1	3	4	6	9	11	18	25
	350	0	0	1	1	1	2	3	6	7	10	15	22
	400	0	0	1	1	1	2	3	5	7	9	14	20
	500	0	0	0	1	1	1	2	4	5	7	11	16
	600	0	0	0	1	1	1	1	3	4	6	9	13
	700	0	0	0	1	1	1	1	3	4	5	8	11
	750	0	0	0	0	1	1	1	3	4	5	7	11
	800	0	0	0	0	1	1	1	2	3	4	7	10

LIQUIDTIGHT FLEXIBLE METALLIC CONDUIT - MAXIMUM NUMBER OF CONDUCTORS

TYPE	SIZE	1/2	3/4	1	1 1/4	1 1/2	2	2 1/2	3	3 1/2	4
		| LFMC Trade Size (Inches)									
RHH, RHW, RHW-2	14	4	7	12	21	27	44	66	102	133	173
	12	3	6	10	17	22	36	55	84	110	144
	10	3	5	8	14	18	29	44	68	89	116
	8	1	2	4	7	9	15	23	36	46	61
	6	1	1	3	6	7	12	18	28	37	48
	4	1	1	2	4	6	9	14	22	29	38
	3	1	1	1	4	5	8	13	19	25	33
	2	1	1	1	3	4	7	11	17	22	29
	1	0	1	1	1	3	5	7	11	14	19
	1/0	0	1	1	1	2	4	6	10	13	16
	2/0	0	1	1	1	1	3	5	8	11	14
	3/0	0	0	1	1	1	3	4	7	9	12
	4/0	0	0	1	1	1	2	4	6	8	10
	250	0	0	0	1	1	1	3	4	6	8
	300	0	0	0	1	1	1	2	4	5	7
	350	0	0	0	1	1	1	2	3	5	6
	400	0	0	0	1	1	1	1	3	4	6
	500	0	0	0	1	1	1	1	3	4	5
	600	0	0	0	0	1	1	1	2	3	4
	700	0	0	0	0	1	1	1	1	3	3
	750	0	0	0	0	0	1	1	1	2	3
	800	0	0	0	0	0	1	1	1	2	3
	1000	0	0	0	0	0	1	1	1	1	3
TW, THHW, THW, THW-2	14	9	15	25	44	57	93	140	215	280	365
	12	7	12	19	33	43	71	108	165	215	280
	10	5	9	14	25	32	53	80	123	160	209
	8	3	5	8	14	18	29	44	68	89	116
RHH*, RHW*, RHW-2*	14	6	10	16	29	38	62	93	143	186	243
	12	5	8	13	23	30	50	75	115	149	195
	10	3	6	10	18	23	39	58	89	117	152
	8	1	4	6	11	14	23	35	53	70	91
RHH*, RHW*, RHW-2*, TW, THW, THHW, THW-2	6	1	3	5	8	11	18	27	41	53	70
	4	1	1	3	6	8	13	20	30	40	52
	3	1	1	3	5	7	11	17	26	34	44
	2	1	1	2	4	6	9	14	22	29	38
	1	1	1	1	3	4	7	10	15	20	26
	1/0	0	1	1	2	3	6	8	13	17	23
	2/0	0	1	1	2	3	5	7	11	15	19
NOTE:	3/0	0	1	1	1	2	4	6	9	12	16
(*) Denotes	4/0	0	0	1	1	1	3	5	8	10	13
Types Minus	250	0	0	1	1	1	3	4	6	8	11
Outer	300	0	0	1	1	1	2	3	5	7	9
Covering	350	0	0	0	1	1	1	3	5	6	8
	400	0	0	0	1	1	1	3	4	6	7
	500	0	0	0	1	1	1	2	3	5	6
	600	0	0	0	1	1	1	1	3	4	5
	700	0	0	0	0	1	1	1	2	3	4
	750	0	0	0	0	1	1	1	2	3	4
	800	0	0	0	0	1	1	1	2	3	4
	1000	0	0	0	0	0	1	1	1	2	3
THHN, THWN, THWN-2	14	13	22	36	63	81	133	201	308	401	523
	12	9	16	26	46	59	97	146	225	292	381
	10	6	10	16	29	37	61	92	141	184	240
	8	3	6	9	16	21	35	53	81	106	138
	6	2	4	7	12	15	25	38	59	76	100
	4	1	2	4	7	9	15	23	36	47	61
	3	1	1	3	6	8	13	20	30	40	52
	2	1	1	3	5	7	11	17	26	33	44
	1	1	1	1	4	5	8	12	19	25	32

LIQUIDTIGHT FLEXIBLE METALLIC CONDUIT - MAXIMUM NUMBER OF CONDUCTORS

TYPE	SIZE	LFMC Trade Size (Inches)									
		1/2	3/4	1	1¼	1½	2	2½	3	3½	4
THHN,	1/0	1	1	1	3	4	7	10	16	21	27
THWN,	2/0	0	1	1	2	3	6	8	13	17	23
THWN-2	3/0	0	1	1	1	3	5	7	11	14	19
(cont.)	4/0	0	1	1	1	2	4	6	9	12	15
	250	0	0	1	1	1	3	5	7	10	12
	300	0	0	1	1	1	3	4	6	8	11
	350	0	0	1	1	1	2	3	5	7	9
	400	0	0	0	1	1	1	3	5	6	8
	500	0	0	0	1	1	1	2	4	5	7
	600	0	0	0	1	1	1	1	3	4	6
	700	0	0	0	1	1	1	1	3	4	5
	750	0	0	0	0	1	1	1	3	3	5
	800	0	0	0	0	1	1	1	2	3	4
	1000	0	0	0	0	0	1	1	1	3	3
FEP, FEPB,	14	12	21	35	61	79	129	195	299	389	507
PFA, PFAH,	12	9	15	25	44	57	94	142	218	284	370
TFE	10	6	11	18	32	41	68	102	156	203	266
	8	3	6	10	18	23	39	58	89	117	152
	6	2	4	7	13	17	27	41	64	83	108
	4	1	3	5	9	12	19	29	44	58	75
	3	1	2	4	7	10	16	24	37	48	63
	2	1	1	3	6	8	13	20	30	40	52
PFA, PFAH, TFE	1	1	1	2	4	5	9	14	21	28	36
PFA, PFAH,	1/0	1	1	1	3	4	7	11	18	23	30
TFE, Z	2/0	1	1	1	3	4	6	9	14	19	25
	3/0	0	1	1	2	3	5	8	12	16	20
	4/0	0	1	1	1	2	4	6	10	13	17
Z	14	20	26	42	73	95	156	235	360	469	611
	12	14	18	30	52	67	111	167	255	332	434
	10	8	11	18	32	41	68	102	156	203	266
	8	5	7	11	20	26	43	64	99	129	168
	6	4	5	8	14	18	30	45	69	90	118
	4	2	3	5	9	12	20	31	48	62	81
	3	2	2	4	7	9	15	23	35	45	59
	2	1	1	3	6	7	12	19	29	38	49
	1	1	1	2	5	6	10	15	23	30	40
XHH,	14	9	15	25	44	57	93	140	215	280	365
XHHW,	12	7	12	19	33	43	71	108	165	215	280
XHHW-2,	10	5	9	14	25	32	53	80	123	160	209
ZW	8	3	5	8	14	18	29	44	68	89	116
	6	1	3	6	10	13	22	33	50	66	86
	4	1	2	4	7	9	16	24	36	48	62
	3	1	1	3	6	8	13	20	31	40	52
	2	1	1	3	5	7	11	17	26	34	44
XHH,	1	1	1	1	4	5	8	12	19	25	33
XHHW,	1/0	1	1	1	3	4	7	10	16	21	28
XHHW-2	2/0	0	1	1	2	3	6	9	13	17	23
	3/0	0	1	1	1	3	5	7	11	14	19
	4/0	0	1	1	1	2	4	6	9	12	16
	250	0	0	1	1	1	3	5	7	10	13
	300	0	0	1	1	1	3	4	6	8	11
	350	0	0	1	1	1	2	3	5	7	10
	400	0	0	0	1	1	1	3	5	6	8
	500	0	0	0	1	1	1	2	4	5	7
	600	0	0	0	1	1	1	1	3	4	6
	700	0	0	0	1	1	1	1	3	4	5
	750	0	0	0	0	1	1	1	3	3	5
	800	0	0	0	0	1	1	1	2	3	4

NON-METALLIC TUBING - MAXIMUM NUMBER OF CONDUCTORS

TYPE	SIZE	ENT Trade Size (Inches)					
		1/2	3/4	1	1 1/4	1 1/2	2
RHH, RHW,	14	3	6	10	19	26	43
RHW-2	12	2	5	9	16	22	36
	10	1	4	7	13	17	29
	8	1	1	3	6	9	15
	6	1	1	3	5	7	12
	4	1	1	2	4	6	9
	3	1	1	1	3	5	8
	2	0	1	1	3	4	7
	1	0	1	1	1	3	5
	1/0	0	0	1	1	2	4
	2/0	0	0	1	1	1	3
	3/0	0	0	1	1	1	3
	4/0	0	0	1	1	1	2
	250	0	0	0	1	1	1
	300	0	0	0	1	1	1
	350	0	0	0	1	1	1
	400	0	0	0	1	1	1
	500	0	0	0	0	1	1
	600	0	0	0	0	1	1
	700	0	0	0	0	0	1
	750	0	0	0	0	0	1
	800	0	0	0	0	0	1
	1000	0	0	0	0	0	1
TW, THHW,	14	7	13	22	40	55	92
THW, THW-2	12	5	10	17	31	42	71
	10	4	7	13	23	32	52
	8	1	4	7	13	17	29
RHH*, RHW*,	14	4	8	15	27	37	61
RHW-2*	12	3	7	12	21	29	49
	10	3	5	9	17	23	38
	8	1	3	5	10	14	23
RHH*, RHW*,	6	1	2	4	7	10	17
RHW-2*, TW,	4	1	1	3	5	8	13
THW, THHW,	3	1	1	2	5	7	11
THW-2	2	1	1	2	4	6	9
	1	0	1	1	3	4	6
	1/0	0	1	1	2	3	5
	2/0	0	1	1	1	3	5
NOTE:	3/0	0	0	1	1	2	4
(*) Denotes	4/0	0	0	1	1	1	3
Types Minus	250	0	0	1	1	1	2
Outer	300	0	0	0	1	1	2
Covering	350	0	0	0	1	1	1
	400	0	0	0	1	1	1
	500	0	0	0	1	1	1
	600	0	0	0	0	1	1
	700	0	0	0	0	1	1
	750	0	0	0	0	1	1
	800	0	0	0	0	1	1
	1000	0	0	0	0	0	1
THHN,	14	10	18	32	58	80	132
THWN,	12	7	13	23	42	58	96
THWN-2	10	4	8	15	26	36	60
	8	2	5	8	15	21	35
	6	1	3	6	11	15	25
	4	1	1	4	7	9	15
	3	1	1	3	5	8	13
	2	1	1	2	5	6	11
	1	1	1	1	3	5	8

NON-METALLIC TUBING - MAXIMUM NUMBER OF CONDUCTORS

TYPE	SIZE	ENT Trade Size (Inches)					
		1/2	3/4	1	1 1/4	1 1/2	2
THHN, THWN, THWN-2 (cont.)	1/0	0	1	1	3	4	7
	2/0	0	1	1	2	3	5
	3/0	0	1	1	1	3	4
	4/0	0	0	1	1	2	4
	250	0	0	1	1	1	3
	300	0	0	1	1	1	2
	350	0	0	0	1	1	2
	400	0	0	0	1	1	1
	500	0	0	0	1	1	1
	600	0	0	0	0	1	1
	700	0	0	0	0	1	1
	750	0	0	0	0	1	1
	800	0	0	0	0	1	1
	1000	0	0	0	0	0	1
FEP, FEPB, PFA, PFAH, TFE	14	10	18	31	56	77	128
	12	7	13	23	41	56	93
	10	5	9	16	29	40	67
	8	3	5	9	17	23	38
	6	1	4	6	12	16	27
	4	1	2	4	8	11	19
	3	1	1	4	7	9	16
	2	1	1	3	5	8	13
PFA, PFAH, TFE	1	1	1	1	4	5	9
PFA, PFAH, TFE, Z	1/0	0	1	1	3	4	7
	2/0	0	1	1	2	4	6
	3/0	0	1	1	1	3	5
	4/0	0	1	1	1	2	4
Z	14	12	22	38	68	93	154
	12	8	15	27	48	66	109
	10	5	9	16	29	40	67
	8	3	6	10	18	25	42
	6	1	4	7	13	18	30
	4	1	3	5	9	12	20
	3	1	1	3	6	9	15
	2	1	1	3	5	7	12
	1	1	1	2	4	6	10
XHH, XHHW, XHHW-2, ZW	14	7	13	22	40	55	92
	12	5	10	17	31	42	71
	10	4	7	13	23	32	52
	8	1	4	7	13	17	29
	6	1	3	5	9	13	21
	4	1	1	4	7	9	15
	3	1	1	3	6	8	13
	2	1	1	2	5	6	11
XHH, XHHW, XHHW-2	1	1	1	1	3	5	8
	1/0	0	1	1	3	4	7
	2/0	0	1	1	2	3	6
	3/0	0	1	1	1	3	5
	4/0	0	0	1	1	2	4
	250	0	0	1	1	1	3
	300	0	0	1	1	1	3
	350	0	0	1	1	1	2
	400	0	0	0	1	1	1
	500	0	0	0	1	1	1
	600	0	0	0	1	1	1
	700	0	0	0	0	1	1
	750	0	0	0	0	1	1
	800	0	0	0	0	1	1

ELECTRICAL METALLIC TUBING - MAXIMUM NUMBER OF CONDUCTORS

TYPE	SIZE	1/2	3/4	1	1 1/4	1 1/2	2	2 1/2	3	3 1/2	4
							EMT Trade Size (Inches)				
RHH, RHW,	14	4	7	11	20	27	46	80	120	157	201
RHW-2	12	3	6	9	17	23	38	66	100	131	167
	10	2	5	8	13	18	30	53	81	105	135
	8	1	2	4	7	9	16	28	42	55	70
	6	1	1	3	5	8	13	22	34	44	56
	4	1	1	2	4	6	10	17	26	34	44
	3	1	1	1	4	5	9	15	23	30	38
	2	1	1	1	3	4	7	13	20	26	33
	1	0	1	1	1	3	5	9	13	17	22
	1/0	0	1	1	1	2	4	7	11	15	19
	2/0	0	1	1	1	2	4	6	10	13	17
	3/0	0	0	1	1	1	3	5	8	11	14
	4/0	0	0	1	1	1	3	5	7	9	12
	250	0	0	0	1	1	1	3	5	7	9
	300	0	0	0	1	1	1	3	5	6	8
	350	0	0	0	1	1	1	3	4	6	7
	400	0	0	0	1	1	1	2	4	5	7
	500	0	0	0	0	1	1	2	3	4	6
	600	0	0	0	0	1	1	1	3	4	5
	700	0	0	0	0	0	1	1	2	3	4
	750	0	0	0	0	0	1	1	2	3	4
	800	0	0	0	0	0	1	1	2	3	4
	1000	0	0	0	0	0	1	1	1	2	3
TW, THHW,	14	8	15	25	43	58	96	168	254	332	424
THW, THW-2	12	6	11	19	33	45	74	129	195	255	326
	10	5	8	14	24	33	55	96	145	190	243
	8	2	5	8	13	18	30	53	81	105	135
RHH*, RHW*,	14	6	10	16	28	39	64	112	169	221	282
RHW-2*	12	4	8	13	23	31	51	90	136	177	227
	10	3	6	10	18	24	40	70	106	138	177
	8	1	4	6	10	14	24	42	63	83	106
RHH* RHW*,	6	1	3	4	8	11	18	32	48	63	81
RHW-2*, TW,	4	1	1	3	6	8	13	24	36	47	60
THW, THHW,	3	1	1	3	5	7	12	20	31	40	52
THW-2	2	1	1	2	4	6	10	17	26	34	44
	1	1	1	1	3	4	7	12	18	24	31
	1/0	0	1	1	2	3	6	10	16	20	26
	2/0	0	1	1	1	3	5	9	13	17	22
NOTE:	3/0	0	1	1	1	2	4	7	11	15	19
(*) Denotes	4/0	0	0	1	1	1	3	6	9	12	16
Types Minus	250	0	0	1	1	1	3	5	7	10	13
Outer	300	0	0	1	1	1	2	4	6	8	11
Covering	350	0	0	0	1	1	1	4	6	7	10
	400	0	0	0	1	1	1	3	5	7	9
	500	0	0	0	1	1	1	3	4	6	7
	600	0	0	0	1	1	1	2	3	4	6
	700	0	0	0	0	1	1	1	3	4	5
	750	0	0	0	0	1	1	1	3	4	5
	800	0	0	0	0	1	1	1	3	3	5
	1000	0	0	0	0	0	1	1	2	3	4
THHN,	14	12	22	35	61	84	138	241	364	476	608
THWN,	12	9	16	26	45	61	101	176	266	347	443
THWN-2	10	5	10	16	28	38	63	111	167	219	279
	8	3	6	9	16	22	36	64	96	126	161
	6	2	4	7	12	16	26	46	69	91	116
	4	1	2	4	7	10	16	28	43	56	71
	3	1	1	3	6	8	13	24	36	47	60
	2	1	1	3	5	7	11	20	30	40	51
	1	1	1	1	4	5	8	15	22	29	37

ELECTRICAL METALLIC TUBING - MAXIMUM NUMBER OF CONDUCTORS

TYPE	SIZE	1/2	3/4	1	11/4	11/2	2	21/2	3	31/2	4
THHN,	1/0	1	1	1	3	4	7	12	19	25	32
THWN,	2/0	0	1	1	2	3	6	10	16	20	26
THWN-2	3/0	0	1	1	1	3	5	8	13	17	22
(cont.)	4/0	0	1	1	1	2	4	7	11	14	18
	250	0	0	1	1	1	3	6	9	11	15
	300	0	0	1	1	1	3	5	7	10	13
	350	0	0	1	1	1	2	4	6	9	11
	400	0	0	0	1	1	1	4	6	8	10
	500	0	0	0	1	1	1	3	5	6	8
	600	0	0	0	1	1	1	2	4	5	7
	700	0	0	0	1	1	1	2	3	4	6
	750	0	0	0	0	1	1	1	3	4	5
	800	0	0	0	0	1	1	1	3	4	5
	1000	0	0	0	0	0	1	1	2	3	4
FEP, FEPB,	14	12	21	34	60	81	134	234	354	462	590
PFA, PFAH,	12	9	15	25	43	59	98	171	258	337	430
TFE	10	6	11	18	31	42	70	122	185	241	309
	8	3	6	10	18	24	40	70	106	138	177
	6	2	4	7	12	17	28	50	75	98	126
	4	1	3	5	9	12	20	35	53	69	88
	3	1	2	4	7	10	16	29	44	57	73
	2	1	1	3	6	8	13	24	36	47	60
PFA, PFAH, TFE	1	1	1	2	4	6	9	16	25	33	42
PFA, PFAH,	1/0	1	1	1	3	5	8	14	21	27	35
TFE, Z	2/0	0	1	1	3	4	6	11	17	22	29
	3/0	0	1	1	2	3	5	9	14	18	24
	4/0	0	1	1	1	2	4	8	11	15	19
Z	14	14	25	41	72	98	161	282	426	556	711
	12	10	18	29	51	69	114	200	302	394	504
	10	6	11	18	31	42	70	122	185	241	309
	8	4	7	11	20	27	44	77	117	153	195
	6	3	5	8	14	19	31	54	82	107	137
	4	1	3	5	9	13	21	37	56	74	94
	3	1	2	4	7	9	15	27	41	54	69
	2	1	1	3	6	8	13	22	34	45	57
	1	1	1	2	4	6	10	18	28	36	46
XHH,	14	8	15	25	43	58	96	168	254	332	424
XHHW,	12	6	11	19	33	45	74	129	195	255	326
XHHW-2,	10	5	8	14	24	33	55	96	145	190	243
ZW	8	2	5	8	13	18	30	53	81	105	135
	6	1	3	6	10	14	22	39	60	78	100
	4	1	2	4	7	10	16	28	43	56	72
	3	1	1	3	6	8	14	24	36	48	61
	2	1	1	3	5	7	11	20	31	40	51
XHH,	1	1	1	1	4	5	8	15	23	30	38
XHHW,	1/0	1	1	1	3	4	7	13	19	25	32
XHHW-2	2/0	0	1	1	2	3	6	10	16	21	27
	3/0	0	1	1	1	3	5	9	13	17	22
	4/0	0	1	1	1	2	4	7	11	14	18
	250	0	0	1	1	1	3	6	9	12	15
	300	0	0	1	1	1	3	5	8	10	13
	350	0	0	1	1	1	2	4	7	9	11
	400	0	0	0	1	1	1	4	6	8	10
	500	0	0	0	1	1	1	3	5	6	8
	600	0	0	0	1	1	1	2	4	5	6
	700	0	0	0	0	1	1	2	3	4	6
	750	0	0	0	0	1	1	1	3	4	5
	800	0	0	0	0	1	1	1	3	4	5

FLEXIBLE METALLIC CONDUIT - MAXIMUM NUMBER OF CONDUCTORS

TYPE	SIZE	FMC Trade Size (Inches)									
		1/2	3/4	1	1 1/4	1 1/2	2	2 1/2	3	3 1/2	4
RHH, RHW, RHW-2	14	4	7	11	17	25	44	67	96	131	171
	12	3	6	9	14	21	37	55	80	109	142
	10	3	5	7	11	17	30	45	64	88	115
	8	1	2	4	6	9	15	23	34	46	60
	6	1	1	3	5	7	12	19	27	37	48
	4	1	1	2	4	5	10	14	21	29	37
	3	1	1	1	3	5	8	13	18	25	33
	2	1	1	1	3	4	7	11	16	22	28
	1	0	1	1	1	2	5	7	10	14	19
	1/0	0	1	1	1	2	4	6	9	12	16
	2/0	0	1	1	1	1	3	5	8	11	14
	3/0	0	0	1	1	1	3	5	7	9	12
	4/0	0	0	1	1	1	2	4	6	8	10
	250	0	0	0	1	1	1	3	4	6	8
	300	0	0	0	1	1	1	2	4	5	7
	350	0	0	0	1	1	1	2	3	5	6
	400	0	0	0	0	1	1	1	3	4	6
	500	0	0	0	0	1	1	1	3	4	5
	600	0	0	0	0	1	1	1	2	3	4
	700	0	0	0	0	0	1	1	1	3	3
	750	0	0	0	0	0	1	1	1	2	3
	800	0	0	0	0	0	1	1	1	2	3
	1000	0	0	0	0	0	1	1	1	1	3
TW, THHW, THW, THW-2	14	9	15	23	36	53	94	141	203	277	361
	12	7	11	18	28	41	72	108	156	212	277
	10	5	8	13	21	30	54	81	116	158	207
	8	3	5	7	11	17	30	45	64	88	115
RHH*, RHW*, RHW-2*	14	6	10	15	24	35	62	94	135	184	240
	12	5	8	12	19	28	50	75	108	148	193
	10	4	6	10	15	22	39	59	85	115	151
	8	1	4	6	9	13	23	35	51	69	90
RHH*, RHW*, RHW-2*, TW, THW, THHW, THW-2	6	1	3	4	7	10	18	27	39	53	69
	4	1	1	3	5	7	13	20	29	39	51
	3	1	1	3	4	6	11	17	25	34	44
	2	1	1	2	4	5	10	14	21	29	37
	1	1	1	1	2	4	7	10	15	20	26
	1/0	0	1	1	1	3	6	9	12	17	22
	2/0	0	1	1	1	3	5	7	10	14	19
NOTE:	3/0	0	1	1	1	2	4	6	9	12	16
(*) Denotes	4/0	0	0	1	1	1	3	5	7	10	13
Types Minus	250	0	0	1	1	1	3	4	6	8	11
Outer	300	0	0	1	1	1	2	3	5	7	9
Covering	350	0	0	0	1	1	1	3	4	6	8
	400	0	0	0	1	1	1	3	4	6	7
	500	0	0	0	1	1	1	2	3	5	6
	600	0	0	0	0	1	1	1	3	4	5
	700	0	0	0	0	1	1	1	2	3	4
	750	0	0	0	0	1	1	1	2	3	4
	800	0	0	0	0	1	1	1	1	3	4
	1000	0	0	0	0	0	1	1	1	2	3
THHN, THWN, THWN-2	14	13	22	33	52	76	134	202	291	396	518
	12	9	16	24	38	56	98	147	212	289	378
	10	6	10	15	24	35	62	93	134	182	238
	8	3	6	9	14	20	35	53	77	105	137
	6	2	4	6	10	14	25	38	55	76	99
	4	1	2	4	6	9	16	24	34	46	61
	3	1	1	3	5	7	13	20	29	39	51
	2	1	1	3	4	6	11	17	24	33	43
	1	1	1	1	3	4	8	12	18	24	32

FLEXIBLE METALLIC CONDUIT-MAXIMUM NUMBER OF CONDUCTORS

TYPE	AWG/kcmil	FMC Trade Size (Inches)									
		1/2	3/4	1	1 1/4	1 1/2	2	2 1/2	3	3 1/2	4
THHN, THWN, THWN-2 (cont.)	1/0	1	1	1	2	4	7	10	15	20	27
	2/0	0	1	1	1	3	6	9	12	17	22
	3/0	0	1	1	1	2	5	7	10	14	18
	4/0	0	1	1	1	1	4	6	8	12	15
	250	0	0	1	1	1	3	5	7	9	12
	300	0	0	1	1	1	3	4	6	8	11
	350	0	0	1	1	1	2	3	5	7	9
	400	0	0	0	1	1	1	3	5	6	8
	500	0	0	0	1	1	1	2	4	5	7
	600	0	0	0	0	1	1	1	3	4	5
	700	0	0	0	0	1	1	1	3	4	5
	750	0	0	0	0	1	1	1	2	3	4
	800	0	0	0	0	1	1	1	2	3	4
	1000	0	0	0	0	0	1	1	1	1	3
FEP, FEPB, PFA, PFAH, TFE	14	12	21	32	51	74	130	196	282	385	502
	12	9	15	24	37	54	95	143	206	281	367
	10	6	11	17	26	39	68	103	148	201	263
	8	4	6	10	15	22	39	59	85	115	151
	6	2	4	7	11	16	28	42	60	82	107
	4	1	3	5	7	11	19	29	42	57	75
	3	1	2	4	6	9	16	24	35	48	62
	2	1	1	3	5	7	13	20	29	39	51
PFA, PFAH, TFE	1	1	1	2	3	5	9	14	20	27	36
PFA, PFAH, TFE, Z	1/0	1	1	2	3	4	8	11	17	23	30
	2/0	1	1	1	2	3	6	9	14	19	24
	3/0	0	1	1	1	3	5	8	11	15	20
	4/0	0	1	1	1	2	4	6	9	13	16
Z	14	15	25	39	61	89	157	236	340	463	605
	12	11	18	28	43	63	111	168	241	329	429
	10	6	11	17	26	39	68	103	148	201	263
	8	4	7	11	17	24	43	65	93	127	166
	6	3	5	7	12	17	30	45	65	89	117
	4	1	3	5	8	12	21	31	45	61	80
	3	1	2	4	6	8	15	23	33	45	58
	2	1	1	3	5	7	12	19	27	37	49
	1	1	1	2	4	6	10	15	22	30	39
XHH, XHHW, XHHW-2, ZW	14	9	15	23	36	53	94	141	203	277	361
	12	7	11	18	28	41	72	108	156	212	277
	10	5	8	13	21	30	54	81	116	158	207
	8	3	5	7	11	17	30	45	64	88	115
	6	1	3	5	8	12	22	33	48	65	85
	4	1	2	4	6	9	16	24	34	47	61
	3	1	1	3	5	7	13	20	29	40	52
	2	1	1	1	4	6	11	17	24	33	44
XHH, XHHW, XHHW-2	1	1	1	1	3	5	8	13	18	25	32
	1/0	1	1	1	2	4	7	10	15	21	27
	2/0	0	1	1	2	3	6	9	13	17	23
	3/0	0	1	1	1	3	5	7	10	14	19
	4/0	0	1	1	1	2	4	6	9	12	15
	250	0	0	1	1	1	3	5	7	10	13
	300	0	0	1	1	1	3	4	6	8	11
	350	0	0	1	1	1	2	4	5	7	9
	400	0	0	0	1	1	1	3	5	6	8
	500	0	0	0	1	1	1	3	4	5	7
	600	0	0	0	0	1	1	1	3	4	5
	700	0	0	0	0	1	1	1	3	4	5
	750	0	0	0	0	1	1	1	2	3	4
	800	0	0	0	0	1	1	1	2	3	4

RIGID PVC SCHEDULE 40 CONDUIT - MAXIMUM NUMBER OF CONDUCTORS

TYPE	SIZE	PVC 40 Trade Size (Inches)											
		1/2	3/4	1	11/4	11/2	2	21/2	3	31/2	4	5	6
RHH, RHW, RHW-2	14	4	7	11	20	27	45	64	99	133	171	269	390
	12	3	5	9	16	22	37	53	82	110	142	224	323
	10	2	4	7	13	18	30	43	66	89	115	181	261
	8	1	2	4	7	9	15	22	35	46	60	94	137
	6	1	1	3	5	7	12	18	28	37	48	76	109
	4	1	1	2	4	6	10	14	22	29	37	59	85
	3	1	1	1	4	5	8	12	19	25	33	52	75
	2	1	1	1	3	4	7	10	16	22	28	45	65
	1	1	1	1	1	3	5	7	11	14	19	29	43
	1/0	0	1	1	1	2	4	6	9	13	16	26	37
	2/0	0	0	1	1	1	3	5	8	11	14	22	32
	3/0	0	0	1	1	1	3	4	7	9	12	19	28
	4/0	0	0	1	1	1	2	4	6	8	10	16	24
	250	0	0	0	1	1	1	3	4	6	8	12	18
	300	0	0	0	1	1	1	2	4	5	7	11	16
	350	0	0	0	1	1	1	2	3	5	6	10	14
	400	0	0	0	1	1	1	1	3	4	6	9	13
	500	0	0	0	0	1	1	1	3	4	5	8	11
	600	0	0	0	0	1	1	1	2	3	4	6	9
	700	0	0	0	0	0	1	1	1	3	3	6	8
	750	0	0	0	0	0	1	1	1	2	3	5	8
	800	0	0	0	0	0	1	1	1	2	3	5	7
	1000	0	0	0	0	0	1	1	1	1	3	4	6
TW, THHW, THW, THW-2	14	8	14	24	42	57	94	135	209	280	361	568	822
	12	6	11	18	32	44	72	103	160	215	277	436	631
	10	4	8	13	24	32	54	77	119	160	206	325	470
	8	2	4	7	13	18	30	43	66	89	115	181	261
RHH*, RHW*, RHW-2*	14	5	9	16	28	38	63	90	139	186	240	378	546
	12	4	8	12	22	30	50	72	112	150	193	304	439
	10	3	6	10	17	24	39	56	87	117	150	237	343
	8	1	3	6	10	14	23	33	52	70	90	142	205
RHH*, RHW*, RHW-2*, TW, THHW, THW, THW-2 **NOTE:** (*) Denotes Types Minus Outer Covering	6	1	2	4	8	11	18	26	40	53	69	109	157
	4	1	1	3	6	8	13	19	30	40	51	81	117
	3	1	1	3	5	7	11	16	25	34	44	69	100
	2	1	1	2	4	6	10	14	22	29	37	59	85
	1	0	1	1	3	4	7	10	15	20	26	41	60
	1/0	0	1	1	2	3	6	8	13	17	22	35	51
	2/0	0	1	1	1	3	5	7	11	15	19	30	43
	3/0	0	1	1	1	2	4	6	9	12	16	25	36
	4/0	0	0	1	1	1	3	5	8	10	13	21	30
	250	0	0	1	1	1	3	4	6	8	11	17	25
	300	0	0	1	1	1	2	3	5	7	9	15	21
	350	0	0	0	1	1	1	3	5	6	8	13	19
	400	0	0	0	1	1	1	3	4	6	7	12	17
	500	0	0	0	1	1	1	2	3	5	6	10	14
	600	0	0	0	1	1	1	1	3	4	5	8	11
	700	0	0	0	0	1	1	1	3	4	4	7	10
	750	0	0	0	0	1	1	1	2	3	4	6	10
	800	0	0	0	0	1	1	1	2	3	4	6	9
	1000	0	0	0	0	0	1	1	1	2	3	5	7
THHN, THWN, THWN-2	14	11	21	34	60	82	135	193	299	401	517	815	1178
	12	8	15	25	43	59	99	141	218	293	377	594	859
	10	5	9	15	27	37	62	89	137	184	238	374	541
	8	3	5	9	16	21	36	51	79	106	137	216	312
	6	1	4	6	11	15	26	37	57	77	99	156	225
	4	1	2	4	7	9	16	22	35	47	61	96	138
	3	1	1	3	6	8	13	19	30	40	51	81	117
	2	1	1	3	5	7	11	16	25	33	43	68	98
	1	1	1	1	3	5	8	12	18	25	32	50	73

RIGID PVC SCHEDULE 40 CONDUIT-MAXIMUM NUMBER OF CONDUCTORS

TYPE	SIZE	PVC 40 Trade Size (Inches)											
		1/2	3/4	1	1 1/4	1 1/2	2	2 1/2	3	3 1/2	4	5	6
THHN,	1/0	1	1	1	3	4	7	10	15	21	27	42	61
THWN,	2/0	0	1	1	2	3	6	8	13	17	22	35	51
THWN-2	3/0	0	1	1	1	3	5	7	11	14	18	29	42
(cont.)	4/0	0	1	1	1	2	4	6	9	12	15	24	35
	250	0	0	1	1	1	3	4	7	10	12	20	28
	300	0	0	1	1	1	3	4	6	8	11	17	24
	350	0	0	1	1	1	2	3	5	7	9	15	21
	400	0	0	0	1	1	1	3	5	6	8	13	19
	500	0	0	0	1	1	1	2	4	5	7	11	16
	600	0	0	0	1	1	1	1	3	4	5	9	13
	700	0	0	0	0	1	1	1	3	4	5	8	11
	750	0	0	0	0	1	1	1	2	3	4	7	11
	800	0	0	0	0	1	1	1	2	3	4	7	10
	1000	0	0	0	0	0	1	1	1	3	3	6	8
FEP, FEPB,	14	11	20	33	58	79	131	188	290	389	502	790	1142
PFA, PFAH,	12	8	15	24	42	58	96	137	212	284	366	577	834
TFE	10	6	10	17	30	41	69	98	152	204	263	414	598
	8	3	6	10	17	24	39	56	87	117	150	237	343
	6	2	4	7	12	17	28	40	62	83	107	169	244
	4	1	3	5	8	12	19	28	43	58	75	118	170
	3	1	2	4	7	10	16	23	36	48	62	98	142
	2	1	1	3	6	8	13	19	30	40	51	81	117
PFA, PFAH, TFE	1	1	1	2	4	5	9	13	20	28	36	56	81
PFA, PFAH,	1/0	1	1	1	3	4	8	11	17	23	30	47	68
TFE, Z	2/0	0	1	1	3	4	6	9	14	19	24	39	56
	3/0	0	1	1	2	3	5	7	12	16	20	32	46
	4/0	0	1	1	1	2	4	6	9	13	16	26	38
Z	14	13	24	40	70	95	158	226	350	469	605	952	1376
	12	9	17	28	49	68	112	160	248	333	429	675	976
	10	6	10	17	30	41	69	98	152	204	263	414	598
	8	3	6	11	19	26	43	62	96	129	166	261	378
	6	2	4	7	13	18	30	43	67	90	116	184	265
	4	1	3	5	9	12	21	30	46	62	80	126	183
	3	1	2	4	6	9	15	22	34	45	58	92	133
	2	1	1	3	5	7	12	18	28	38	49	77	111
	1	1	1	2	4	6	10	14	23	30	39	62	90
XHH,	14	8	14	24	42	57	94	135	209	280	361	568	822
XHHW,	12	6	11	18	32	44	72	103	160	215	277	436	631
XHHW-2,	10	4	8	13	24	32	54	77	119	160	206	325	470
ZW	8	2	4	7	13	18	30	43	66	89	115	181	261
	6	1	3	5	10	13	22	32	49	66	85	134	193
	4	1	2	4	7	9	16	23	35	48	61	97	140
	3	1	1	3	6	8	13	19	30	40	52	82	118
	2	1	1	3	5	7	11	16	25	34	44	69	99
XHH,	1	1	1	1	3	5	8	12	19	25	32	51	74
XHHW	1/0	1	1	1	3	4	7	10	16	21	27	43	62
XHHW-2	2/0	0	1	1	2	3	6	8	13	17	23	36	52
	3/0	0	1	1	1	3	5	7	11	14	19	30	43
	4/0	0	1	1	1	2	4	6	9	12	15	24	35
	250	0	0	1	1	1	3	4	6	10	13	20	29
	300	0	0	1	1	1	3	4	6	8	11	17	25
	350	0	0	1	1	1	2	3	5	7	9	15	22
	400	0	0	0	1	1	1	3	5	6	8	13	19
	500	0	0	0	1	1	1	2	4	5	7	11	16
	600	0	0	0	1	1	1	1	3	4	5	9	13
	700	0	0	0	0	1	1	1	3	4	5	8	11
	750	0	0	0	0	1	1	1	3	4	4	7	11
	800	0	0	0	0	1	1	1	2	3	4	7	10

RIGID PVC SCHEDULE 80 CONDUIT-MAXIMUM NUMBER OF CONDUCTORS

TYPE	SIZE	1/2	3/4	1	1 1/4	1 1/2	2	2 1/2	3	3 1/2	4	5	6
						PVC 80 Trade Size (Inches)							
RHH, RHW, RHW-2	14	3	5	9	17	23	39	56	88	118	153	243	349
	12	2	4	7	14	19	32	46	73	98	127	202	290
	10	1	3	6	11	15	26	37	59	79	103	163	234
	8	1	1	3	6	8	13	19	31	41	54	85	122
	6	1	1	2	4	6	11	16	24	33	43	68	98
	4	1	1	1	3	5	8	12	19	26	33	53	77
	3	0	1	1	3	4	7	11	17	23	29	47	67
	2	0	1	1	3	4	6	9	14	20	25	41	58
	1	0	1	1	1	2	4	6	9	13	17	27	38
	1/0	0	0	1	1	1	3	5	8	11	15	23	33
	2/0	0	0	1	1	1	3	4	7	10	13	20	29
	3/0	0	0	1	1	1	3	4	6	8	11	17	25
	4/0	0	0	0	1	1	2	3	5	7	9	15	21
	250	0	0	0	1	1	1	2	4	5	7	11	16
	300	0	0	0	1	1	1	2	3	5	6	10	14
	350	0	0	0	1	1	1	1	3	4	5	9	13
	400	0	0	0	1	1	1	1	3	4	5	8	12
	500	0	0	0	0	1	1	1	2	3	4	7	10
	600	0	0	0	0	1	0	1	1	3	3	6	8
	700	0	0	0	0	0	1	1	1	2	3	5	7
	750	0	0	0	0	0	1	1	1	2	3	5	7
	800	0	0	0	0	0	1	1	1	2	3	4	7
	1000	0	0	0	0	0	1	1	1	1	2	4	5
TW, THHW, THW, THW-2	14	6	11	20	35	49	82	118	185	250	324	514	736
	12	5	9	15	27	38	63	91	142	192	248	394	565
	10	3	6	11	20	28	47	67	106	143	185	294	421
	8	1	3	6	11	15	26	37	59	79	103	163	234
RHH*, RHW*, RHW-2*	14	4	8	13	23	32	55	79	123	166	215	341	490
	12	3	6	10	19	26	44	63	99	133	173	274	394
	10	2	5	8	15	20	34	49	77	104	135	214	307
	8	1	3	5	9	12	20	29	46	62	81	128	184
RHH*, RHW*, RHW-2*, TW, THW, THHW, THW-2	6	1	1	3	7	9	16	22	35	48	62	98	141
	4	1	1	3	5	7	12	17	26	35	46	73	105
	3	1	1	2	4	6	10	14	22	30	39	63	90
	2	1	1	1	3	5	8	12	19	26	33	53	77
	1	0	1	1	2	3	6	8	13	18	23	37	54
	1/0	0	1	1	1	3	5	7	11	15	20	32	46
	2/0	0	1	1	1	2	4	6	10	13	17	27	39
NOTE: (*) Denotes Types Minus Outer Covering	3/0	0	0	1	1	1	3	5	8	11	14	23	33
	4/0	0	0	1	1	1	3	4	7	9	12	19	27
	250	0	0	0	1	1	2	3	5	7	9	15	22
	300	0	0	0	1	1	1	3	5	6	8	13	19
	350	0	0	0	1	1	1	2	4	6	7	12	17
	400	0	0	0	1	1	1	2	4	5	7	10	15
	500	0	0	0	1	1	1	1	3	4	5	9	13
	600	0	0	0	0	1	1	1	2	3	4	7	10
	700	0	0	0	0	1	1	1	2	3	4	6	9
	750	0	0	0	0	0	1	1	1	3	4	6	8
	800	0	0	0	0	0	1	1	1	3	3	6	8
	1000	0	0	0	0	0	1	1	1	2	3	5	7
THHN, THWN, THWN-2	14	9	17	28	51	70	118	170	265	358	464	736	1055
	12	6	12	20	37	51	86	124	193	261	338	537	770
	10	4	7	13	23	32	54	78	122	164	213	338	485
	8	2	4	7	13	18	31	45	70	95	123	195	279
	6	1	3	5	9	13	22	32	51	68	89	141	202
	4	1	1	3	6	8	14	20	31	42	54	86	124
	3	1	1	3	5	7	12	17	26	35	46	73	105
	2	1	1	2	4	6	10	14	22	30	39	61	88
	1	0	1	1	3	4	7	10	16	22	29	45	65

RIGID PVC SCHEDULE 80 CONDUIT-MAXIMUM NUMBER OF CONDUCTORS

TYPE	SIZE	PVC 80 Trade Size (Inches)											
		1/2	3/4	1	1 1/4	1 1/2	2	2 1/2	3	3 1/2	4	5	6
THHN,	1/0	0	1	1	2	3	6	9	14	18	24	38	55
THWN,	2/0	0	1	1	1	3	5	7	11	15	20	32	46
THWN-2	3/0	0	1	1	1	2	4	6	9	13	17	26	38
(cont.)	4/0	0	0	1	1	1	3	5	8	10	14	22	31
	250	0	0	1	1	1	2	4	6	8	11	18	25
	300	0	0	1	1	1	2	3	5	7	9	15	22
	350	0	0	0	1	1	1	3	5	6	8	13	19
	400	0	0	0	1	1	1	3	4	6	7	12	17
	500	0	0	0	1	1	1	2	3	5	6	10	14
	600	0	0	0	0	1	1	1	3	4	5	8	12
	700	0	0	0	0	1	1	1	2	3	4	7	10
	750	0	0	0	0	1	1	1	2	3	4	7	9
	800	0	0	0	0	1	1	1	2	3	4	6	9
	1000	0	0	0	0	1	1	1	2	3	5	5	7
FEP, FEPB,	14	8	16	27	49	68	115	164	257	347	450	714	1024
PFA, PFAH,	12	6	12	20	36	50	84	120	188	253	328	521	747
TFE	10	4	8	14	26	36	60	86	135	182	235	374	536
	8	2	5	8	15	20	34	49	77	104	135	214	307
	6	1	3	6	10	14	24	35	55	74	96	152	218
	4	1	2	4	7	10	17	24	38	52	67	106	153
	3	1	1	3	6	8	14	20	32	43	56	89	127
	2	1	1	3	5	7	12	17	26	35	46	73	105
PFA, PFAH, TFE	1	1	1	1	3	5	8	11	18	25	32	51	73
PFA, PFAH,	1/0	0	1	1	3	4	7	10	15	20	27	42	61
TFE, Z	2/0	0	1	1	2	3	5	8	12	17	22	35	50
	3/0	0	1	1	1	2	4	6	10	14	18	29	41
	4/0	0	0	1	1	1	3	5	8	11	15	24	34
Z	14	10	19	33	59	82	138	198	310	418	542	860	1233
	12	7	14	23	42	58	98	141	220	297	385	610	875
	10	4	8	14	26	36	60	86	135	182	235	374	536
	8	3	5	9	16	22	38	54	85	115	149	236	339
	6	2	4	6	11	16	26	38	60	81	104	166	238
	4	1	2	4	8	11	18	26	41	55	72	114	164
	3	1	1	3	5	8	13	19	30	40	52	83	119
	2	1	1	2	5	6	11	16	25	33	43	69	99
	1	0	1	2	4	5	9	13	20	27	35	56	80
XHH,	14	6	11	20	35	49	82	118	185	250	324	514	736
XHHW,	12	5	9	15	27	38	63	91	142	192	248	394	565
XHHW-2,	10	3	6	11	20	28	47	67	106	143	185	294	421
ZW	8	1	3	6	11	15	26	37	59	79	103	163	234
	6	1	2	4	8	11	19	28	43	59	76	121	173
	4	1	1	3	6	8	14	20	31	42	55	87	125
	3	1	1	3	5	7	12	17	26	36	47	74	106
	2	1	1	2	4	6	10	14	22	30	39	62	89
XHH, XHHW,	1	1	1	1	3	4	7	10	16	22	29	46	66
XHHW-2	1/0	0	1	1	2	3	6	9	14	19	24	39	56
	2/0	0	1	1	1	3	5	7	11	16	20	32	46
	3/0	0	1	1	1	2	4	6	9	13	17	27	38
	4/0	0	0	1	1	1	3	5	8	11	14	22	32
	250	0	0	1	1	1	3	4	6	9	11	18	26
	300	0	0	1	1	1	2	3	5	7	10	15	22
	350	0	0	0	1	1	1	3	5	6	8	14	20
	400	0	0	0	1	1	1	3	4	6	7	12	17
	500	0	0	0	1	1	1	2	3	5	6	10	14
	600	0	0	0	0	1	1	1	3	4	5	8	11
	700	0	0	0	0	1	1	1	2	3	4	7	10
	750	0	0	0	0	1	1	1	2	3	4	6	9
	800	0	0	0	0	1	1	1	1	3	4	6	9

WIRE DATA
STANDARD STRANDED CONDUCTORS

Size kcmil	Number of Wires in the Strand								
	7	19	37	49	61	91	127	169	217
	Diameter in Inches of Each Wire in the Strand								
2000	.5345	.3243	.2325	.202	.181	.1482	.1255	.1088	.096
1750	.5000	.3034	.2175	.189	.1694	.1386	.1157	.1003	.0898
1500	.4629	.2810	.2013	.175	.1568	.1284	.1087	.0942	.0831
1250	.4226	.2565	.1838	.1507	.1431	.1172	.0992	.086	.0759
1000	.378	.2294	.1644	.1429	.1285	.1048	.0887	.0769	.0678
950	.3684	.2236	.1602	.1392	.1248	.1021	.0864	.075	.0662
900	.3586	.2176	.1559	.1355	.1215	.0994	.0841	.073	.0644
850	.3484	.2115	.1516	.1317	.1181	.0966	.0818	.0709	.0626
800	.338	.205	.147	.1278	.1145	.0937	.0793	.0687	.0607
750	.3273	.1986	.1424	.1237	.1109	.0908	.0768	.0666	.0588
700	.3163	.1919	.1375	.1195	.1071	.0883	.0742	.0644	.0568
650	.3047	.1850	.1325	.1152	.1032	.0845	.0716	.0620	.0547
600	.2928	.1778	.1273	.1107	.0992	.0812	.0687	.0596	.0526
550	.2803	.1701	.1219	.106	.0950	.0777	.0658	.0570	.0503
500	.2672	.1622	.1162	.101	.0905	.0741	.0627	.0544	.048
450	.2535	.1539	.1103	.0958	.0859	.0703	.0595	.0516	.0455
400	.2391	.1451	.1040	.0904	.081	.0663	.0561	.0487	.0429
350	.2236	.1357	.0973	.0845	.0757	.0620	.0526	.0455	.0401
300	.207	.1257	.0901	.0783	.0701	.0573	.0486	.0421	.0372
250	.189	.1147	.0822	.0714	.064	.0524	.0444	.0384	.0340
Size AWG									
0000	.1736	.1055	.0756	.0657	.0589	.0482	.0408	–	–
000	.1548	.0940	.0673	.0586	.0525	.0429	.0363	–	–
00	.1378	.0836	.0599	.0521	.0467	.0382	.0323	–	–
0	.1228	.0746	.0534	.0464	.0416	.0340	.0288	–	–
1	.1093	.0663	.0475	.0413	.0370	.0303	.0252	–	–
2	.0973	.0592	.0423	.0369	.0329	.0269	.0228	–	–
3	.0867	.0526	.0377	.0327	.0294	.0240	.0203	–	–
4	.0772	.0468	.0335	.0291	.0261	.0214	.0179	–	–
6	.0612	.0372	.0266	.0231	.0207	.0169	.0143	–	–
8	.0485	.0293	.0211	.0184	.0164	.0135	.0114	–	–
10	.0385	.0223	.0168	–	.0129	.0106	.0090	–	–

COPPER WIRE SPECIFICATIONS

Wire Size (AWG)	Area in Circular Mils	Diameter in Mils (1000th In.)	Diameter Millimeters	Weight in lbs. per 1000 Feet
40	9.9	3.1	.080	.0200
38	15.7	4	.101	.0476
36	25	5	.127	.0757
34	39.8	6.3	.160	.120
32	63.2	8	.202	.191
30	101	10	.255	.304
28	160	12.6	.321	.484
26	254	15.9	.405	.769
24	404	20.1	.511	1.22
22	642	25.3	.644	1.94
20	1,020	32	.812	3.09
18	1,620	40	1.024	4.92
16	2,580	51	1.291	7.82
14	4,110	64	1.628	12.4
12	6,530	81	2.053	19.8
10	10,400	102	2.588	31.4
8	16,500	128	3.264	50
6	26,300	162	4.115	79.5
4	41,700	204	5.189	126
3	52,600	229	5.827	159
2	66,400	258	6.544	201
1	83,700	289	7.348	253
0	106,000	325	8.255	319
00	133,000	365	9.271	403
000	168,000	410	10.414	508
0000	212,000	460	11.684	641

COMPARATIVE WEIGHTS OF COPPER AND ALUMINUM CONDUCTORS/Lbs. per 1,000 Ft.

Size AWG or MCM	Bare - Solid			Bare - Stranded		
	Cu.	Al.	Diff.	Cu.	Al.	Diff.
18	4.92	1.49	3.43	5.02	1.53	3.49
16	7.82	2.38	5.44	7.97	2.43	5.54
14	12.43	3.78	8.65	12.68	3.86	8.82
12	19.77	6.01	13.76	20.16	6.13	14.03
10	31.43	9.55	21.88	32.06	9.75	22.31
8	50.0	15.2	34.8	51.0	15.5	35.5
6	79.5	24.2	55.3	81.0	24.6	56.4
4	126.4	38.4	88.0	128.9	39.2	89.7
2	200.9	60.8	140.1	204.9	62.3	142.6
1	253.3	77.0	176.3	258.4	78.6	179.8
1/0	319.5	97.2	222.3	325.8	99.1	226.7
2/0	402.8	122.6	280.2	410.9	124.9	286.0
3/0	507.9	154.6	353.3	518.1	157.5	360.6
4/0	640.5	194.9	445.6	653.3	198.6	454.7
250				771.9	234.7	537.2
300				926.3	281.6	644.7
350				1081.	328.6	752.
400				1235.	375.5	859.
450				1389.	422.4	967.
500				1544.	469.4	1075.
600				1853.	563.	1290.
700				2161.	657.	1504.
750				2316.	704.	1612.
800				2470.	751.	1719.
900				2779.	845.	1934.
1000				3088.	939.	2149.
1250				3859.	1173.	2686.
1500				4631.	1410.	3221.
1750				5403.	1645.	3758.
2000				6175.	1880.	4295.

COPPER BUS-BAR DATA
Sizes, Weights and Resistances

Thickness (Inches)	Width (Inches)	Wts. per Ft. at .3213 Lbs. per Cubic In.	Area in Square In.	Ohms per Ft. at 8.341 per Sq. Mil. Ft.	Capacity in Amperes
1/16	1/2	.1205	.0313	.00026691	30
1/16	3/4	.1807	.0469	.00017790	50
1/16	1	.2410	.0625	.00013344	60
1/16	1-1/2	.3615	.0938	.00008897	90
1/8	1/2	.2410	.0625	.00013344	75
1/8	3/4	.3615	.0938	.00008897	90
1/8	1	.4820	.125	.00006672	125
1/8	1-1/2	.7230	.1875	.00004448	200
1/8	2	.9640	.25	.00003336	250
1/4	3/4	.7230	.1875	.00004448	185
1/4	1	.9640	.25	.00003336	250
1/4	1-1/4	1.205	.3125	.00002669	315
1/4	1-1/2	1.446	.375	.00002224	375
1/4	1-3/4	1.687	.4375	.00001906	435
1/4	2	1.928	.5	.00001668	500
1/4	2-1/4	2.169	.5625	.00001482	565
1/4	2-1/2	2.410	.625	.00001334	630
1/2	3/4	1.446	.375	.00002224	370
1/2	1	1.928	.500	.00001668	500
1/2	1-1/4	2.410	.625	.00001334	625
1/2	1-1/2	2.892	.750	.00001112	750
1/2	1-3/4	3.374	.875	.00000953	875
1/2	2	3.856	1.	.00000834	1000
1/2	2-1/4	4.338	1.125	.00000741	1185
1/2	2-1/2	4.820	1.25	.00000667	1250
1/2	2-3/4	5.304	1.375	.00000606	1375
1/2	3	5.784	1.500	.00000556	1500
1/2	3-1/4	6.266	1.625	.00000513	1625
1/2	3-1/2	6.748	1.750	.00000475	1750
1/2	3-3/4	7.23	1.875	.00000444	1875
1/2	4	7.712	2.000	.00000417	2000
3/4	1	2.892	.750	.00001112	750
3/4	1-1/2	4.338	1.125	.00000741	1125
3/4	2	5.784	1.500	.00000556	1500
3/4	2-1/2	7.23	1.875	.00000444	1875
3/4	3	8.676	2.250	.00000370	2250
3/4	3-1/2	10.122	2.625	.00000317	2650
3/4	4	11.568	3.000	.00000278	3000

Carrying capacity is figured at 1,000 amperes per square inch.

VOLTAGE DROP USING OHMS LAW

Voltage drop is the amount of voltage lost over the length of a piece of wire. Voltage drop is defined as a function of wire resistance. Efficiency will be decreased if the voltage drop exceeds 3%. An electrical motor, for example, may have a shorter lifespan with a drop greater than 3% because the efficiency of the load may be decreased and operating temperature increased.

One way to calculate voltage drop is to use <u>OHM's LAW</u>. The definition is as follows: Voltage equals current (Amperes) times resistance (Ohms).

A working model using the following criteria will calculate the voltage drop in a power line. The line is 300 feet long using 12 gauge copper wire powering an 800 Watt light fixture from a 120 Volt supply.

The resistance of #12 wire is 1.62 Ohms per 1000 Feet @ 77° F
or 0.00162 Ohms/Foot

The current is 800 Watts/120 Volts = 6.67 Amps
The resistance of the line = 300 Feet x .00162 Ohms/Foot
= 0.486 Ohms

Voltage Drop = 6.67 Amps x 0.486 Ohms = 3.24 Volts

The Voltage Drop in percentage = 3.24 Volts/120 Volts = 2.7%

CALCULATING BRANCH CIRCUIT
VOLTAGE DROP IN PERCENT

$$\% \ V_D = \frac{V_{NL} - V_{FL}}{V_{FL}} \times 100$$

WHERE

$\% \ V_D$ = percent voltage drop

V_{NL} = no-load voltage drop (Volts)

V_{FL} = full-load voltage drop (Volts)

VOLTAGE DROP FORMULAS

The NEC® recommends a maximum 3% voltage drop for either the branch circuit or the feeder.

Single-Phase:

$$VD = \frac{2 \times R \times I \times L}{CM}$$

Three-Phase:

$$VD = \frac{1.732 \times R \times I \times L}{CM}$$

VD = Volts (voltage drop of the circuit)

R = 12.9 Ohms/Copper or 21.2 Ohms/Aluminum (resistance constants for a 1,000 circular mils conductor that is 1,000 feet long, at an operating temperature of 75° C.)

I = Amps (load at 100 percent)

L = Feet (length of circuit from load to power supply)

CM = Circular-Mils (conductor wire size)

2 = Single-Phase Constant

1.732 = Three-Phase Constant

CONDUCTOR LENGTH/VOLTAGE DROP

Voltage drop can be reduced by limiting the length of the conductors.

Single-Phase:

$$L = \frac{CM \times VD}{2 \times R \times I}$$

Three-Phase:

$$L = \frac{CM \times VD}{1.732 \times R \times I}$$

CONDUCTOR SIZE/VOLTAGE DROP

Increase the size of the conductor to decrease the voltage drop of circuit (reduce its resistance).

Single-Phase:

$$CM = \frac{2 \times R \times I \times L}{VD}$$

Three-Phase:

$$CM = \frac{1.732 \times R \times I \times L}{VD}$$

VOLTAGE DROP AMPERE-FEET
Copper Conductors, 70°C Copper Temp.,
600 V Class Single Conductor Cables in Steel Conduit

Conductor Size AWG or MCM	DC Circuits 1% Drop on 120 V	Maximum circuit ampere-feet without exceeding specified percentage voltage drop, various circuit voltages and power factors.				
		60 Cycle AC Circuits				
		1% Drop, 1.00 P.F.		3% Drop, 0.85 P.F.		
		120 V 1-Phase	208 V 3-Phase	115 V 1-Phase	208 V 3-Phase	220 V 3-Phase
14	191	191	382	623	1,300	1.380
12	305	305	612	998	2,080	2,200
10	484	484	968	1,580	3,280	3,470
8	770	770	1,540	2,450	5,110	5,410
6	1,200	1,190	2,380	3,800	7,850	8,310
4	1,900	1,890	3,780	5,750	12,000	12,700
2	3,030	2,970	5,950	8,620	18,000	19,100
1	3,820	3,710	7,430	10,400	21,800	23,100
1/0	4,820	4,560	9,120	12,500	26,000	27,500
2/0	6,060	5,610	11,200	14,800	30,800	32,700
3/0	7,650	6,940	13,900	16,900	35,200	37,100
4/0	9,760	8,520	17,100	20,200	42,100	44,600
250	11,500	9,930	19,800	22,300	46,500	49,300
300	13,800	11,500	23,100	24,800	52,000	55,000
350	16,100	13,200	26,300	27,000	56,200	59,500
400	18,400	14,000	28,100	28,700	60,000	63,500
500	22,700	16,400	32,800	31,800	66,400	70,300
750	34,600	20,400	40,800	36,700	76,600	81,200

NOTE: Length to be used is the distance from point of supply to load.

WIRE LENGTH vs WIRE SIZE (MAX. VOLTAGE DROP)

Max. Wire Feet @ 240 Volts, Single-Phase, 2% Max. Voltage Drop

3/0	2/0	1/0	#2	#4	Watts	Amps
180	-	-	-	-	48,000	200
240	190	185	-	-	36,000	150
360	280	230	-	-	24,000	100
440	365	290	180	-	19,200	80
520	415	330	205	130	16,800	70
600	485	385	240	150	14,400	60
720	580	460	290	180	12,000	50
880	725	575	360	230	5,600	40
1,200	970	770	485	300	7,200	30
1,440	1,100	920	580	365	6,000	25

#6	#8	#10	#12	#14	Watts	Amps
105	-	-	-	-	12,000	50
130	90	-	-	-	5,600	40
175	120	75	-	-	7,200	30
210	144	90	-	-	6,000	25
265	180	110	70	-	4,800	20
350	240	150	95	60	3,600	15
525	360	225	140	90	2,400	10
1,020	720	455	285	180	1,200	5

Max. Wire Feet @ 120 Volts, Single-Phase, 2% Max. Voltage Drop

3/0	2/0	1/0	#2	#4	Watts	Amps
230	180	144	90	-	9,600	80
260	205	165	105	65	8,400	70
305	240	190	120	76	7,200	60
360	290	230	145	90	6,000	50
440	360	290	175	115	4,800	40
600	490	385	240	150	3,600	30
720	580	460	290	180	3,000	25
900	725	575	365	230	2,400	20
1,200	965	770	485	305	1,800	15

#6	#8	#10	#12	#14	Watts	Amps
57	-	-	-	-	6,000	50
72	45	-	-	-	4,800	40
95	60	38	-	-	3,600	30
115	72	45	-	-	3,000	25
140	90	57	36	-	2,400	20
190	120	75	47	30	1,800	15
285	180	115	70	45	1,200	10
575	360	225	140	90	600	5

COPPER WIRE RESISTANCE

Wire Size (AWG)	at 77°F Feet per OHM	at 149°F Feet per OHM	at 77°F OHMS per 1,000 Feet	at 149°F OHMS per 1,000 Feet
40	.93	.81	1,070	1,230
38	1.5	1.3	673	776
36	2.4	2.0	423	488
34	3.8	3.3	266	307
32	6.0	5.2	167	193
30	9.5	8.3	105	121
28	15.1	13.1	66.2	76.4
26	24.0	20.8	41.6	48.0
24	38.2	33.1	26.2	30.2
22	60.6	52.6	16.5	19.0
20	96.2	84.0	10.4	11.9
18	153.6	133.2	6.51	7.51
16	244.5	211.4	4.09	4.73
14	387.6	336.7	2.58	2.97
12	617.3	534.8	1.62	1.87
10	980.4	847.5	1.02	1.18
8	1,560	1,353	.641	.739
6	2,481	2,151	.403	.465
4	3,953	3,425	.253	.292
3	4,975	4,310	.201	.232
2	6,289	5,435	.159	.184
1	7,936	6,849	.126	.146
0	10,000	8,621	.100	.116
00	12,658	10,870	.079	.092
000	15,873	13,699	.063	.073
0000	20,000	17,544	.050	.057

VOLTAGE DROP TABLE

Conductor Size	DC	Volts Drop Per 1000 Ampere-Feet						
		AC System						
		LOAD POWER FACTOR in Percent						
		100	95	90	85	80	75	70

For DC circuit or single phase, 60 cycle, 2-wire system or 3-wire system with balanced load. Copper conductors, 70°C copper temperature. 600 V class single-conductor cables in steel conduit.

Conductor Size	DC	100	95	90	85	80	75	70
14	6.29	6.29	6.06	5.78	5.54	5.26	4.97	4.74
12	3.93	3.93	3.81	3.64	3.46	3.29	3.13	2.95
10	2.48	2.48	2.44	2.31	2.19	2.08	1.96	1.85
8	1.56	1.56	1.51	1.47	1.41	1.34	1.27	1.20
6	0.999	1.011	0.987	0.953	0.918	0.872	0.826	0.774
4	0.631	0.635	0.641	0.624	0.600	0.578	0.554	0.528
2	0.396	0.404	0.418	0.413	0.400	0.386	0.372	0.358
1	0.314	0.323	0.356	0.337	0.330	0.322	0.311	0.300
1/0	0.249	0.263	0.280	0.282	0.277	0.269	0.263	0.255
2/0	0.198	0.214	0.233	0.236	0.233	0.230	0.226	0.222
3/0	0.157	0.173	0.196	0.206	0.204	0.200	0.194	0.188
4/0	0.123	0.141	0.163	0.170	0.171	0.170	0.169	0.166
250 MCM	0.1041	0.121	0.146	0.152	0.155	0.155	0.155	0.154
300 MCM	0.0870	0.1040	0.128	0.135	0.139	0.140	0.141	0.141
350 MCM	0.0746	0.0912	0.117	0.125	0.128	0.131	0.131	0.131
400 MCM	0.0652	0.0855	0.1086	0.117	0.120	0.122	0.124	0.125
500 MCM	0.0528	0.0733	0.0959	0.1040	0.1086	0.111	0.113	0.114
750 MCM	0.0347	0.0589	0.0808	0.0884	0.0940	0.0976	0.0999	0.1020

For 3-phase, 60 cycle, 3-wire or 4-wire balanced system. Copper conductors, 70°C copper temperature, 600 V. class single-conductor cables in steel conduit.

Conductor Size	DC	100	95	90	85	80	75	70
14		5.45	5.25	5.00	4.80	4.55	4.30	4.10
12		3.40	3.30	3.15	3.00	2.85	2.70	2.55
10		2.15	2.10	2.00	1.90	1.80	1.70	1.60
8		1.35	1.31	1.27	1.22	1.16	1.10	1.04
6		0.875	0.855	0.825	0.795	0.755	0.715	0.670
4		0.550	0.555	0.540	0.520	0.500	0.480	0.457
2		0.350	0.362	0.358	0.346	0.334	0.322	0.310
1		0.280	0.308	0.292	0.286	0.279	0.269	0.260
1/0		0.228	0.242	0.244	0.240	0.233	0.228	0.221
2/0		0.185	0.202	0.204	0.202	0.199	0.196	0.192
3/0		0.150	0.170	0.178	0.177	0.173	0.168	0.163
4/0		0.122	0.141	0.147	0.148	0.147	0.146	0.144
250 MCM		0.105	0.126	0.132	0.134	0.134	0.134	0.133
300 MCM		0.0900	0.111	0.117	0.120	0.121	0.122	0.122
350 MCM		0.0790	0.101	0.108	0.111	0.113	0.114	0.114
400 MCM		0.0740	0.0940	0.101	0.104	0.106	0.107	0.108
500 MCM		0.0635	0.0830	0.0900	0.0940	0.0964	0.0974	0.0988
750 MCM		0.0510	0.0700	0.0765	0.0814	0.0845	0.0865	0.0883

NOTE: Length to be used is the distance from point of supply to load, not amount of wire in circuit.

AMPACITIES OF COPPER CONDUCTORS (3)

AMPACITIES OF NOT MORE THAN 3 INSULATED CONDUCTORS RATED 0-2000 VOLTS IN CABLE OR RACEWAY

	IN CABLE, RACEWAY OR EARTH, AMBIENT TEMPERATURE 30°C(86°F)			IN CABLE OR RACEWAY, AMBIENT TEMPERATURE 40°C(104°F)			
WIRE SIZE	**60°C (140°F)**	**75°C (167°F)**	**90°C (194°F)**	**150°C (302°F)**	**200°C (392°F)**	**250°C (482°F)**	**WIRE SIZE**
AWG kcmil	TYPES TW UF	TYPES RHW, THHW, THW, THWN, XHHW, USE, ZW	TYPES TBS, SA, SIS, FEP, MI FEPB, ZW-2, RHH, RHW-2, THHN, THHW, THW-2, XHH, USE-2, THWN-2 XHHW, XHHW-2.	TYPE Z	TYPES FEP, FEPB, PFA	TYPES PFAH, TFE NICKEL OR NICKEL-COATED COPPER	AWG kcmil
14*	20	20	25	34	36	39	14
12*	25	25	30	43	45	54	12
10*	30	35	40	55	60	73	10
8	40	50	55	76	83	93	8
6	55	65	75	96	110	117	6
4	70	85	95	120	125	148	4
3	85	100	110	143	152	166	3
2	95	115	130	160	171	191	2
1	110	130	150	186	197	215	1
1/0	125	150	170	215	229	244	1/0
2/0	145	175	195	251	260	273	2/0
3/0	165	200	225	288	297	308	3/0
4/0	195	230	260	332	346	361	4/0
250	215	255	290				250

TEMPERATURE CORRECTION FACTORS

Ambient Temp.°C	FOR OTHER THAN 30°C			FOR OTHER THAN 40°C			Ambient Temp.°C
	MULTIPLY THE AMPACITIES ABOVE BY THE FACTORS BELOW						
21-25	1.08	1.05	1.04	.95	.97	.98	41-50
26-30	1.00	1.00	1.00	.90	.94	.95	51-60
31-35	.91	.94	.96	.85	.90	.93	61-70
36-40	.82	.88	.91	.80	.87	.90	71-80
41-45	.71	.82	.87	.74	.83	.87	81-90
46-50	.58	.75	.82	.67	.79	.85	91-100
51-55	.41	.67	.76	.52	.71	.79	101-120
56-60		.58	.71	.30	.61	.72	121-140
61-70		.33	.58		.50	.65	141-160
71-80			.41		.35	.58	161-180
						.49	181-200
						.35	201-225

*Unless specifically permitted by the NEC®, overcurrent protection for copper conductors shall not exceed 15 amps for No. 14 AWG, 20 amps for No. 12 AWG and 30 amps for No. 10 AWG.

AMPACITIES OF COPPER CONDUCTORS (3)

AMPACITIES OF NOT MORE THAN 3 INSULATED CONDUCTORS RATED 0-2000 VOLTS IN CABLE OR RACEWAY

	IN CABLE, RACEWAY OR EARTH, AMBIENT TEMPERATURE 30°C(86°F)			IN CABLE OR RACEWAY, AMBIENT TEMPERATURE 40°C(104°F)			
WIRE SIZE	60°C (140°F)	75°C (167°F)	90°C (194°F)	150°C (302°F)	200°C (392°F)	250°C (482°F)	WIRE SIZE
kcmil	TYPES TW, UF	TYPES RHW, THHW, THW, THWN, XHHW, USE, ZW	TYPES TBS, SA, SIS, FEP, MI FEPB, ZW-2, RHH, RHW-2, THHN, THHW, THW-2, XHH, USE-2, THWN-2 XHHW, XHHW-2.	TYPE Z	TYPES FEP, FEPB, PFA	TYPES PFAH, TFE NICKEL OR NICKEL-COATED COPPER	kcmil
300	240	285	320	—	—	—	300
350	260	310	350	—	—	—	350
400	280	335	380	—	—	—	400
500	320	380	430	—	—	—	500
600	355	420	475	—	—	—	600
700	385	460	520	—	—	—	700
750	400	475	535	—	—	—	750
800	410	490	555	—	—	—	800
900	435	520	585	—	—	—	900
1000	455	545	615	—	—	—	1000
1250	495	590	665	—	—	—	1250
1500	520	625	705	—	—	—	1500
1750	545	650	735	—	—	—	1750
2000	560	665	750	—	—	—	2000

TEMPERATURE CORRECTION FACTORS

Ambient Temp.°C	FOR OTHER THAN 30°C MULTIPLY THE AMPACITIES ABOVE BY THE FACTORS BELOW						
21-25	1.08	1.05	1.04				
26-30	1.00	1.00	1.00				
31-35	.91	.94	.96				
36-40	.82	.88	.91				
41-45	.71	.82	.87				
46-50	.58	.75	.82				
51-55	.41	.67	.76				
56-60	—	.58	.71				
61-70	—	.33	.58				
71-80	—	—	.41				

AMPACITIES OF COPPER CONDUCTORS (1)

AMPACITIES OF SINGLE INSULATED CONDUCTORS RATED 0-2000 VOLTS IN FREE AIR

WIRE SIZE	AMBIENT TEMPERATURE 30°C(86°F)			AMBIENT TEMPERATURE 40°C(104°F)			BARE CONDUCTORS WITH MAX. TEMP. 80°C (176°F)
	60°C (140°F)	75°C (167°F)	90°C (194°F)	150°C (302°F)	200°C (392°F)	250°C (482°F)	
AWG kcmil	TYPES TW UF	TYPES RHW, THHW, THW, THWN, XHHW, ZW	TYPES TBS, SA, SIS, FEP, MI, FEPB, ZW-2, RHH, RHW-2, THHN, THHW, THW-2, XHH, USE-2, THWN-2 XHHW, XHHW-2.	TYPE Z	TYPES FEP, FEPB, PFA	TYPES PFAH, TFE — NICKEL OR NICKEL-COATED COPPER	
14*	25	30	35	46	54	59	
12*	30	35	40	60	68	78	
10*	40	50	55	80	90	107	
8	60	70	80	106	124	142	98
6	80	95	105	155	165	205	124
4	105	125	140	190	220	278	155
3	120	145	165	214	252	327	
2	140	170	190	255	293	381	209
1	165	195	220	293	344	440	
1/0	195	230	260	339	399	532	282
2/0	225	265	300	390	467	591	329
3/0	260	310	350	451	546	708	382
4/0	300	360	405	529	629	830	444
250	340	405	455				494

TEMPERATURE CORRECTION FACTORS

Ambient Temp.°C	FOR OTHER THAN 30°C			FOR OTHER THAN 40°C			Ambient Temp.°C
	MULTIPLY THE AMPACITIES ABOVE BY THE FACTORS BELOW						
21-25	1.08	1.05	1.04	.95	.97	.98	41-50
26-30	1.00	1.00	1.00	.90	.94	.95	51-60
31-35	.91	.94	.96	.85	.90	.93	61-70
36-40	.82	.88	.91	.80	.87	.90	71-80
41-45	.71	.82	.87	.74	.83	.87	81-90
46-50	.58	.75	.82	.67	.79	.85	91-100
51-55	.41	.67	.76	.52	.71	.79	101-120
56-60		.58	.71	.30	.61	.72	121-140
61-70		.33	.58		.50	.65	141-160
71-80			.41		.35	.58	161-180
						.49	181-200
						.35	201-225

*Unless specifically permitted by the NEC®, overcurrent protection for copper conductors shall not exceed 15 amps for No. 14 AWG, 20 amps for No. 12 AWG and 30 amps for No. 10 AWG.

AMPACITIES OF COPPER CONDUCTORS (1)

AMPACITIES OF SINGLE INSULATED CONDUCTORS RATED 0-2000 VOLTS IN FREE AIR

WIRE SIZE	AMBIENT TEMPERATURE 30°C(86°F)			AMBIENT TEMPERATURE 40°C(104°F)			BARE CONDUC-TORS WITH MAX. TEMP. 80°C (176°F)
	60°C (140°F)	75°C (167°F)	90°C (194°F)	150°C (302°F)	200°C (392°F)	250°C (482°F)	
kcmil	TYPES TW, UF	TYPES RHW, THHW, THW, THWN, XHHW, ZW	TYPES TBS, SA, SIS, FEP, MI, FEPB, ZW-2, RHH, RHW-2, THHN, THHW, THW-2, XHH, USE-2, THWN-2 XHHW, XHHW-2.	TYPE Z	TYPES FEP, FEPB, PFA	TYPES PFAH, TFE NICKEL OR NICKEL-COATED COPPER	
300	375	445	505	——	——	——	556
350	420	505	570	——	——	——	
400	455	545	615	——	——	——	
500	515	620	700	——	——	——	773
600	575	690	780	——	——	——	
700	630	755	855	——	——	——	
750	655	785	885	——	——	——	1000
800	680	815	920	——	——	——	
900	730	870	985	——	——	——	
1000	780	935	1055	——	——	——	1193
1250	890	1065	1200	——	——	——	
1500	980	1175	1325	——	——	——	
1750	1070	1280	1445	——	——	——	
2000	1155	1385	1560	——	——	——	

TEMPERATURE CORRECTION FACTORS							
Ambient Temp.°C	FOR OTHER THAN 30°C MULTIPLY THE AMPACITIES ABOVE BY THE FACTORS BELOW						
21-25	1.08	1.05	1.04				
26-30	1.00	1.00	1.00				
31-35	.91	.94	.96				
36-40	.82	.88	.91				
41-45	.71	.82	.87				
46-50	.58	.75	.82				
51-55	.41	.67	.76				
56-60	——	.58	.71				
61-70	——	.33	.58				
71-80	——	——	.41				

AMPACITIES OF ALUMINUM AND COPPER-CLAD ALUMINUM CONDUCTORS (3)

AMPACITIES OF NOT MORE THAN 3 SINGLE INSULATED CONDUCTORS RATED 0-2000 VOLTS IN CABLE OR RACEWAY

WIRE SIZE	IN CABLE, RACEWAY OR EARTH, AMBIENT TEMPERATURE 30°C(86°F)			IN CABLE OR RACEWAY AMBIENT TEMP. 40°C(104°F)	WIRE SIZE
	60°C (140°F)	75°C (167°F)	90°C (194°F)	150°C (302°F)	
AWG kcmil	TYPES TW, UF	TYPES THWN, THHW, XHHW, USE, THW, RHW	TYPES TBS, SA, SIS, THHN, THHW, THW-2, THWN-2, RHH, RHW-2, USE-2, XHH, XHHW, XHHW-2, ZW-2	TYPE Z	AWG kcmil
12*	20	20	25	30	12
10*	25	30	35	44	10
8	30	40	45	57	8
6	40	50	60	75	6
4	55	65	75	94	4
3	65	75	85	109	3
2	75	90	100	124	2
1	85	100	115	145	1
1/0	100	120	135	169	1/0
2/0	115	135	150	198	2/0
3/0	130	155	175	227	3/0
4/0	150	180	205	260	4/0
250	170	205	230		250

TEMPERATURE CORRECTION FACTORS

Ambient Temp.°C	FOR OTHER THAN 30°C			40°C	Ambient Temp.°C
	MULTIPLY THE AMPACITIES ABOVE BY THE FACTORS BELOW				
21-25	1.08	1.05	1.04	.95	41-50
26-30	1.00	1.00	1.00	.90	51-60
31-35	.91	.94	.96	.85	61-70
36-40	.82	.88	.91	.80	71-80
41-45	.71	.82	.87	.74	81-90
46-50	.58	.75	.82	.67	91-100
51-55	.41	.67	.76	.52	101-120
56-60		.58	.71	.30	121-140
61-70		.33	.58		
71-80			.41		

*Unless specifically permitted by the NEC®, overcurrent protection for aluminum and copper-clad aluminum conductors shall not exceed 15 amps for No. 12 AWG and 25 amps for No. 10 AWG.

AMPACITIES OF ALUMINUM AND COPPER-CLAD ALUMINUM CONDUCTORS (3)

AMPACITIES OF NOT MORE THAN 3 SINGLE INSULATED CONDUCTORS RATED 0-2000 VOLTS IN CABLE OR RACEWAY

	IN CABLE, RACEWAY OR EARTH, AMBIENT TEMPERATURE 30°C(86°F)			IN CABLE OR RACEWAY AMBIENT TEMP. 40°C(104°F)	
WIRE SIZE	60°C (140°F)	75°C (167°F)	90°C (194°F)	150°C (302°F)	WIRE SIZE
kcmil	TYPES TW, UF	TYPES THWN, THHW, XHHW, USE, THW, RHW	TYPES TBS, SA, SIS, THHN, THHW, THW-2, THWN-2, RHH, RHW-2, USE-2, XHH, XHHW, XHHW-2, ZW-2	TYPE Z	kcmil
300	190	230	255	—	300
350	210	250	280	—	350
400	225	270	305	—	400
500	260	310	350	—	500
600	285	340	385	—	600
700	310	375	420	—	700
750	320	385	435	—	750
800	330	395	450	—	800
900	355	425	480	—	900
1000	375	445	500	—	1000
1250	405	485	545	—	1250
1500	435	520	585	—	1500
1750	455	545	615	—	1750
2000	470	560	630	—	2000

TEMPERATURE CORRECTION FACTORS			
Ambient Temp.°C	FOR OTHER THAN 30°C MULTIPLY THE AMPACITIES ABOVE BY THE FACTORS BELOW		
21-25	1.08	1.05	1.04
26-30	1.00	1.00	1.00
31-35	.91	.94	.96
36-40	.82	.88	.91
41-45	.71	.82	.87
46-50	.58	.75	.82
51-55	.41	.67	.76
56-60		.58	.71
61-70		.33	.58
71-80			.41

AMPACITIES OF ALUMINUM OR COPPER-CLAD ALUMINUM CONDUCTORS (1)

AMPACITIES OF SINGLE INSULATED CONDUCTORS RATED 0-2000 VOLTS IN FREE AIR

	AMBIENT TEMPERATURE 30°C(86°F)			AMBIENT TEMP. 40°C(104°F)	
WIRE SIZE	60°C (140°F)	75°C (167°F)	90°C (194°F)	150°C (302°F)	WIRE SIZE
AWG kcmil	TYPES TW, UF	TYPES THWN, THHW, XHHW, RHW, THW	TYPES TBS, XHH, RHH, RHW-2, XHHW, USE-2, THHN, SA, THHW, THW-2, XHHW-2, ZW-2, THWN-2, SIS	TYPE Z	AWG kcmil
12*	25	30	35	47	12
10*	35	40	40	63	10
8	45	55	60	83	8
6	60	75	80	112	6
4	80	100	110	148	4
3	95	115	130	170	3
2	110	135	150	198	2
1	130	155	175	228	1
1/0	150	180	205	263	1/0
2/0	175	210	235	305	2/0
3/0	200	240	275	351	3/0
4/0	235	280	315	411	4/0
250	265	315	355		250

TEMPERATURE CORRECTION FACTORS

Ambient Temp.°C	FOR OTHER THAN 30°C			40°C	Ambient Temp.°C
	MULTIPLY THE AMPACITIES ABOVE BY THE FACTORS BELOW				
21-25	1.08	1.05	1.04	.95	41-50
26-30	1.00	1.00	1.00	.90	51-60
31-35	.91	.94	.96	.85	61-70
36-40	.82	.88	.91	.80	71-80
41-45	.71	.82	.87	.74	81-90
46-50	.58	.75	.82	.67	91-100
51-55	.41	.67	.76	.52	101-120
56-60		.58	.71	.30	121-140
61-70		.33	.58		
71-80			.41		

*Unless specifically permitted by the NEC®, overcurrent protection for aluminum and copper-clad aluminum conductors shall not exceed 15 amps for No. 12 AWG and 25 amps for No. 10 AWG.

AMPACITIES OF ALUMINUM OR COPPER-CLAD ALUMINUM CONDUCTORS (1)

AMPACITIES OF SINGLE INSULATED CONDUCTORS RATED 0-2000 VOLTS IN FREE AIR

	AMBIENT TEMPERATURE 30°C (86°F)			AMBIENT TEMP. 40°C (104°F)	
WIRE SIZE	**60°C (140°F)**	**75°C (167°F)**	**90°C (194°F)**	**150°C (302°F)**	**WIRE SIZE**
kcmil	**TYPES** TW, UF	**TYPES** THWN, THHW, XHHW, RHW, THW	**TYPES** TBS, XHH RHH, RHW-2, XHHW, USE-2, THHN, SA, THHW, THW-2, XHHW-2, ZW-2, THWN-2, SIS	**TYPE** Z	kcmil
300	290	350	395	—	300
350	330	395	445	—	350
400	355	425	480	—	400
500	405	485	545	—	500
600	455	540	615	—	600
700	500	595	675	—	700
750	515	620	700	—	750
800	535	645	725	—	800
900	580	700	785	—	900
1000	625	750	845	—	1000
1250	710	855	960	—	1250
1500	795	950	1075	—	1500
1750	875	1050	1185	—	1750
2000	960	1150	1335	—	2000

TEMPERATURE CORRECTION FACTORS				
Ambient Temp.°C	FOR OTHER THAN 30°C MULTIPLY THE AMPACITIES ABOVE BY THE FACTORS BELOW			
21-25	1.08	1.05	1.04	
26-30	1.00	1.00	1.00	
31-35	.91	.94	.96	
36-40	.82	.88	.91	
41-45	.71	.82	.87	
46-50	.58	.75	.82	
51-55	.41	.67	.76	
56-60		.58	.71	
61-70		.33	.58	
71-80			.41	

AMPACITY ADJUSTMENTS FOR 4 OR MORE CONDUCTORS IN A CABLE OR RACEWAY

NUMBER OF CURRENT-CARRYING CONDUCTORS	PERCENT OF VALUES IN AMPACITY CHARTS/ADJUST FOR AMBIENT TEMPERATURE (IF NECESSARY)
4 to 6	80%
7 to 9	70%
10 to 20	50%
21 to 30	45%
31 to 40	40%
41 and above	35%

NOTE: For use with ampacity charts on pages 3–42 through 3–49.

AMPERAGE RATINGS FOR SINGLE-PHASE SERVICE OR FEEDER CONDUCTORS IN NORMAL DWELLING UNITS

COPPER	ALUMINUM OR COPPER-CLAD ALUMINUM	SERVICE OR FEEDER RATING IN AMPS
4 AWG	2 AWG	100
3	1	110
2	1/0	125
1	2/0	150
1/0	3/0	175
2/0	4/0	200
3/0	250 kcmil	225
4/0	300	250
250 kcmil	350	300
350	500	350
400	600	400

MINIMUM SIZE CONDUCTORS FOR GROUNDING RACEWAY AND EQUIPMENT

AMPERAGE RATING OR SETTING OF AUTOMATIC OVERCURRENT DEVICE NOT TO EXCEED	CONDUCTOR SIZE	
	COPPER	ALUMINUM OR COPPER-CLAD ALUMINUM
15	14 AWG	12 AWG
20	12	10
30	10	8
40	10	8
60	10	8
100	8	6
200	6	4
300	4	2
400	3	1
500	2	1/0
600	1	2/0
800	1/0	3/0
1000	2/0	4/0
1200	3/0	250 kcmil
1600	4/0	350
2000	250 kcmil	400
2500	350	600
3000	400	600
4000	500	800
5000	700	1200
6000	800	1200

The equipment grounding conductor shall be sized larger than this table per NEC® installation restrictions in Article 250.

GROUNDING ELECTRODE CONDUCTORS - AC SYSTEMS

SERVICE-ENTRANCE CONDUCTOR OR EQUIVALENT AREA FOR PARALLEL CONDUCTORS		GROUNDING ELECTRODE CONDUCTOR	
COPPER	ALUMINUM OR COPPER-CLAD ALUMINUM	COPPER	ALUMINUM OR COPPER-CLAD ALUMINUM*
2 OR SMALLER	1/0 OR SMALLER	8	6
1 OR 1/0	2/0 OR 3/0 AWG	6	4
2/0 OR 3/0 AWG	4/0 OR 250 kcmil	4	2
OVER 3/0 THRU 350 kcmil	OVER 250 THRU 500 kcmil	2	1/0
OVER 350 kcmil THRU 600 kcmil	OVER 500 kcmil THRU 900 kcmil	1/0	3/0
OVER 600 kcmil THRU 1100 kcmil	OVER 900 kcmil THRU 1750 kcmil	2/0	4/0
OVER 1100 kcmil	OVER 1750 kcmil	3/0	250 kcmil

A) The table above applies to the derived conductors of separately derived AC systems.

B) When multiple sets of service conductors are utilized, the equivalent size of the largest service-entrance conductor shall be determined by the largest sum of the areas of the corresponding conductors of each set.

C) If there are no service-entrance conductors, the grounding electrode conductor size shall be determined by the equivalent size of the largest service-entrance conductor required for the load to be served.

*NOTE: Please refer to NEC® Installation restrictions in Article 250 concerning aluminum and copper-clad aluminum conductors.

CONDUIT AND TUBING - ALLOWABLE AREA DIMENSIONS FOR WIRE COMBINATIONS

TRADE SIZE INCHES	TRADE I.D. INCHES	100% TOTAL AREA SQ. IN.	31% 2 WIRES SQ. IN.	40% OVER 2 WIRES SQ. IN.	53% 1 WIRE SQ. IN.
RIGID METAL CONDUIT					
1/2	0.632	0.314	0.097	0.125	0.166
3/4	0.836	0.549	0.170	0.220	0.291
1	1.063	0.887	0.275	0.355	0.470
1 1/4	1.394	1.526	0.473	0.610	0.809
1 1/2	1.624	2.071	0.642	0.829	1.098
2	2.083	3.408	1.056	1.363	1.806
2 1/2	2.489	4.866	1.508	1.946	2.579
3	3.090	7.499	2.325	3.000	3.974
3 1/2	3.570	10.010	3.103	4.004	5.305
4	4.050	12.882	3.994	5.153	6.828
5	5.073	20.212	6.266	8.085	10.713
6	6.093	29.158	9.039	11.663	15.454
LIQUIDTIGHT FLEXIBLE METAL CONDUIT					
3/8	0.494	0.192	0.059	0.077	0.102
1/2	0.632	0.314	0.097	0.125	0.166
3/4	0.830	0.541	0.168	0.216	0.287
1	1.054	0.873	0.270	0.349	0.462
1 1/4	1.395	1.528	0.474	0.611	0.810
1 1/2	1.588	1.981	0.614	0.792	1.050
2	2.033	3.246	1.006	1.298	1.720
2 1/2	2.493	4.881	1.513	1.953	2.587
3	3.085	7.475	2.317	2.990	3.962
3 1/2	3.520	9.731	3.017	3.893	5.158
4	4.020	12.692	3.935	5.077	6.727
LIQUIDTIGHT FLEXIBLE NONMETALLIC CONDUIT (TYPE LFNC-A)					
3/8	0.495	0.192	0.060	0.077	0.102
1/2	0.630	0.312	0.097	0.125	0.165
3/4	0.825	0.535	0.166	0.214	0.283
1	1.043	0.854	0.265	0.342	0.453
1 1/4	1.383	1.502	0.466	0.601	0.796
1 1/2	1.603	2.018	0.626	0.807	1.070
2	2.063	3.343	1.036	1.337	1.772
LIQUIDTIGHT FLEXIBLE NONMETALLIC CONDUIT (TYPE LFNC-B)					
3/8	0.494	0.192	0.059	0.077	0.102
1/2	0.632	0.314	0.097	0.125	0.166
3/4	0.830	0.541	0.168	0.216	0.287
1	1.054	0.873	0.270	0.349	0.462
1 1/4	1.395	1.528	0.474	0.611	0.810
1 1/2	1.588	1.981	0.614	0.792	1.050
2	2.033	3.246	1.006	1.298	1.720

CONDUIT AND TUBING - ALLOWABLE AREA DIMENSIONS FOR WIRE COMBINATIONS

TRADE SIZE INCHES	TRADE I.D. INCHES	100% TOTAL AREA SQ. IN.	31% 2 WIRES SQ. IN.	40% OVER 2 WIRES SQ. IN.	53% 1 WIRE SQ. IN.
ELECTRICAL NONMETALLIC TUBING					
1/2	0.560	0.246	0.076	0.099	0.131
3/4	0.760	0.454	0.141	0.181	0.240
1	1.000	0.785	0.243	0.314	0.416
1 1/4	1.340	1.410	0.437	0.564	0.747
1 1/2	1.570	1.936	0.600	0.774	1.026
2	2.020	3.205	0.993	1.282	1.699
ELECTRICAL METALLIC TUBING					
1/2	0.622	0.304	0.094	0.122	0.161
3/4	0.824	0.533	0.165	0.213	0.283
1	1.049	0.864	0.268	0.346	0.458
1 1/4	1.380	1.496	0.464	0.598	0.793
1 1/2	1.610	2.036	0.631	0.814	1.079
2	2.067	3.356	1.040	1.342	1.778
2 1/2	2.731	5.858	1.816	2.343	3.105
3	3.356	8.846	2.742	3.538	4.688
3 1/2	3.834	11.545	3.579	4.618	6.119
4	4.334	14.753	4.573	5.901	7.819
INTERMEDIATE METAL CONDUIT					
1/2	0.660	0.342	0.106	0.137	0.181
3/4	0.864	0.586	0.182	0.235	0.311
1	1.105	0.959	0.297	0.384	0.508
1 1/4	1.448	1.647	0.510	0.659	0.873
1 1/2	1.683	2.225	0.690	0.890	1.179
2	2.150	3.630	1.125	1.452	1.924
2 1/2	2.557	5.135	1.592	2.054	2.722
3	3.176	7.922	2.456	3.169	4.199
3 1/2	3.671	10.584	3.281	4.234	5.610
4	4.166	13.631	4.226	5.452	7.224
FLEXIBLE METAL CONDUIT					
3/8	0.384	0.116	0.036	0.046	0.061
1/2	0.635	0.317	0.098	0.127	0.168
3/4	0.824	0.533	0.165	0.213	0.283
1	1.020	0.817	0.253	0.327	0.433
1 1/4	1.275	1.277	0.396	0.511	0.677
1 1/2	1.538	1.858	0.576	0.743	0.985
2	2.040	3.269	1.013	1.307	1.732
2 1/2	2.500	4.909	1.522	1.963	2.602
3	3.000	7.069	2.191	2.827	3.746
3 1/2	3.500	9.621	2.983	3.848	5.099
4	4.000	12.566	3.896	5.027	6.660

CONDUIT AND TUBING - ALLOWABLE AREA DIMENSIONS FOR WIRE COMBINATIONS

TRADE SIZE INCHES	TRADE I.D. INCHES	100% TOTAL AREA SQ. IN.	31% 2 WIRES SQ. IN.	40% OVER 2 WIRES SQ. IN.	53% 1 WIRE SQ. IN.
PVC CONDUIT - TYPE EB					
2	2.221	3.874	1.201	1.550	2.053
3	3.330	8.709	2.700	3.484	4.616
3 1/2	3.804	11.365	3.523	4.546	6.023
4	4.289	14.448	4.479	5.779	7.657
5	5.316	22.195	6.881	8.878	11.763
6	6.336	31.530	9.774	12.612	16.711
RIGID PVC CONDUIT - TYPE A					
1/2	0.700	0.385	0.119	0.154	0.204
3/4	0.910	0.650	0.202	0.260	0.345
1	1.175	1.084	0.336	0.434	0.575
1 1/4	1.500	1.767	0.548	0.707	0.937
1 1/2	1.720	2.324	0.720	0.929	1.231
2	2.155	3.647	1.131	1.459	1.933
2 1/2	2.635	5.453	1.690	2.181	2.890
3	3.230	8.194	2.540	3.278	4.343
3 1/2	3.690	10.694	3.315	4.278	5.668
4	4.180	13.723	4.254	5.489	7.273
RIGID PVC SCHEDULE 40 CONDUIT (HDPE)					
1/2	0.602	0.285	0.088	0.114	0.151
3/4	0.804	0.508	0.157	0.203	0.269
1	1.029	0.832	0.258	0.333	0.441
1 1/4	1.360	1.453	0.450	0.581	0.770
1 1/2	1.590	1.986	0.616	0.794	1.052
2	2.047	3.291	1.020	1.316	1.744
2 1/2	2.445	4.695	1.455	1.878	2.488
3	3.042	7.268	2.253	2.907	3.852
3 1/2	3.521	9.737	3.018	3.895	5.161
4	3.998	12.554	3.892	5.022	6.654
5	5.016	19.761	6.126	7.904	10.473
6	6.031	28.567	8.856	11.427	15.141
RIGID PVC SCHEDULE 80 CONDUIT					
1/2	0.526	0.217	0.067	0.087	0.115
3/4	0.722	0.409	0.127	0.164	0.217
1	0.936	0.688	0.213	0.275	0.365
1 1/4	1.255	1.237	0.383	0.495	0.656
1 1/2	1.476	1.711	0.530	0.684	0.907
2	1.913	2.874	0.891	1.150	1.523
2 1/2	2.290	4.119	1.277	1.647	2.183
3	2.864	6.442	1.997	2.577	3.414
3 1/2	3.326	8.688	2.693	3.475	4.605
4	3.786	11.258	3.490	4.503	5.967
5	4.768	17.855	5.535	7.142	9.463
6	5.709	25.598	7.935	10.239	13.567

DESIGNED DIMENSIONS AND WEIGHTS OF RIGID STEEL CONDUIT

Nominal or trade size of Conduit (Inches)	Inside Diameter (Inches)	Outside Diameter (Inches)	Wall Thickness (Inches)	Length Without Coupling Ft. & Ins.	Minimum Weight of Ten Unit Lengths with Couplings Attached (Pounds)
$1/4$	0.364	0.540	0.088	9–11-$1/2$	38.5
$3/8$	0.493	0.675	0.091	9–11-$1/2$	51.5
$1/2$	0.622	0.840	0.109	9–11-$1/4$	79.0
$3/4$	0.824	1.050	0.113	9–11-$1/4$	105.0
1	1.049	1.315	0.133	9–11	153.0
$1^1/4$	1.380	1.660	0.140	9–11	201.0
$1^1/2$	1.610	1.900	0.145	9–11	249.0
2	2.067	2.375	0.154	9–11	334.0
$2^1/2$	2.469	2.875	0.203	9–10-$1/2$	527.0
3	3.068	3.500	0.216	9–10-$1/2$	690.0
$3^1/2$	3.548	4.000	0.226	9–10-$1/2$	831.0
4	4.026	4.500	0.237	9–10-$1/4$	982.0
5	5.047	5.563	0.258	9–10	1344.0
6	6.065	6.625	0.280	9–10	1770.0

NOTE: The tolerances are:
Length: $\pm$ $1/4$-inch (without coupling)
Outside Diameter $+$ $1/64$-inch or $-$$1/32$ -inch for the $1^1/2$-inch and smaller sizes
$\pm$ 1 percent for the 2-inch and larger sizes
Wall Thickness: $-$ $12^1/2$ percent

DIMENSIONS OF THREADS FOR RIGID STEEL CONDUIT

Nominal or trade size of Conduit (Inches)	Threads per Inch	Pitch Diameter at End of Thread E_0 (Inches) Taper $3/4$ inch per Foot	Length of Thread (Inches)	
			Effective L_2	Over-All L_4
$1/4$	18	0.4774	0.40	0.59
$3/8$	18	0.6120	0.41	0.60
$1/2$	14	0.7584	0.53	0.78
$3/4$	14	0.9677	0.55	0.79
1	11-$1/2$	1.2136	0.68	0.98
$1^1/4$	11-$1/2$	1.5571	0.71	1.01
$1^1/2$	11-$1/2$	1.7961	0.72	1.03
2	11-$1/2$	2.2690	0.76	1.06
2-$1/2$	8	2.7195	1.14	1.57
3	8	3.3406	1.20	1.63
$3^1/2$	8	3.8375	1.25	1.68
4	8	4.3344	1.30	1.73
5	8	5.3907	1.41	1.84
6	8	6.4461	1.51	1.95

NOTE: The tolerances are:
Thread length (L_4, Column 5): $\pm$ 1 thread
Pitch Diameter (Column 3): $\pm$ 1 turn is the maximum variation permitted from the gauging face of the working thread gauges. This is equivalent to $\pm$ 1-1/2 turns from basic dimensions, since a variation of $\pm$1/2 turn from basic dimensions is permitted in working gauges.

DESIGNED DIMENSIONS AND WEIGHTS OF COUPLINGS

Nominal or Trade Size of Conduit (Inches)	Outside Diameter (Inches)	Minimum Length (Inches)	Minimum Weight (Pounds)
1/4	0.719	1-3/16	0.055
3/8	0.875	1-3/16	0.075
1/2	1.010	1-9/16	0.115
3/4	1.250	1-5/8	0.170
1	1.525	2	0.300
1-1/4	1.869	2-1/16	0.370
1-1/2	2.155	2-1/16	0.515
2	2.650	2-1/8	0.671
2-1/2	3.250	3-1/8	1.675
3	3.870	3-1/4	2.085
3-1/2	4.500	3-3/8	3.400
4	4.875	3-1/2	2.839
5	6.000	3-3/4	4.462
6	7.200	4	7.282

NOTE: The tolerances are:
Outside Diameter: −1 percent for the 1-1/4-inch and larger sizes.
−1/64-inch for sizes smaller than 1-1/4-inch.
No limit is placed on the plus tolerances given for this dimension.

DIMENSIONS OF 90-DEGREE ELBOWS AND WEIGHTS OF NIPPLES PER HUNDRED

Nominal or Trade Size of Conduit (Inches)	Elbows		Nipples	
	Minimum Radius to Center of Conduit (Inches)	Minimum Straight Length L_s at Each End (Inches)	A	B
1/4	–	–	–	–
3/8	–	–	–	–
1/2	4	1-1/2	0.065	2
3/4	4-1/2	1-1/2	0.086	4
1	5-3/4	1-7/8	0.125	9
1-1/4	7-1/4	2	0.164	10
1-1/2	8-1/4	2	0.202	11
2	9-1/2	2	0.269	14
2-1/2	10-1/2	3	0.430	60
3	13	3-1/8	0.561	70
3-1/2	15	3-1/4	0.663	90
4	16	3-3/8	0.786	115
5	24	3-5/8	1.060	170
6	30	3-3/4	1.410	200

Each lot of 100 nipples shall weigh not less than the number of pounds determined by the formula:
$$W = 100 \, LA - B$$
Where W = weight of 100 nipples in pounds   L = length of one nipple in inches
A = weight of nipple per inch in pounds
B = weight in pounds, lost in threading 100 nipples

APPROXIMATE SPACING OF CONDUIT BUSHINGS, CHASE NIPPLES AND LOCK NUTS

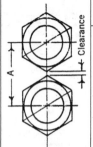

"D" in Simplest Practical Dimension	
Conduit	"D"
1/2	1-1/8
3/4	1-3/8
1	1-5/8
1-1/4	2
1-1/2	2-1/4
2	2-7/8
2-1/2	3-1/2
3	4-5/16
3-1/2	4-7/8

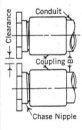

Conduit
Coupling
Chase Nipple

Size of Conduit		Clearance = 1/8"								
		1/2"	3/4'	1"	1-1/4"	1-1/2"	2"	2-1/2"	3"	3-1/2"
1/2"	A	1-1/4	1-3/8	1-1/2	1-11/16	1-13/16	2-1/8	2-7/16	2-3/4	3-1/8
	B	.41	.43	.42	.43	.44	.52	.58	.58	.70
3/4"	A	1-3/8	1-1/2	1-5/8	1-13/16	1-15/16	2-1/4	2-9/16	3	3-1/4
	B	.43	.45	.44	.46	.46	.54	.60	.60	.72
1"	A	1-1/2	1-5/8	1-3/4	1-15/16	2-1/16	2-3/8	2-11/16	3	3-3/8
	B	.43	.45	.44	.46	.46	.54	.60	.60	.73
1-1/4"	A	1-11/16	1-13/16	1-15/16	2-1/8	2-1/4	2-9/16	2-7/8	3-1/4	3-9/16
	B	.43	.46	.46	.46	.47	.55	.62	.67	.74
1-1/2"	A	1-13/16	1-15/16	2-1/16	2-1/4	2-3/8	2-11/16	3	3-3/8	3-11/16
	B	.44	.46	.46.	.47	.47	.56	.62	.67	.74
2"	A	2-1/8	2-1/4	2-3/8	2-9/16	2-11/16	3	3-5/16	3-5/8	4
	B	.52	.54	.53	.55	.56	.63	.63	.69	.82
2-1/2"	A	2-7/16	2-9/16	2-11/16	2-7/8	3	3-5/16	3-5/8	3-15/16	4-5/16
	B	.58	.60	.60	.61	.62	.69	.76	.76	.89
3"	A	2-3/4	3	3	3-1/4	3-3/8	3-5/8	3-15/16	4-7/16	4-3/4
	B	.58	.60	.60	.67	.67	.67	.69	.74	.94	1.00
3-1/2"	A	3-1/8	3-1/4	3-3/8	3-9/16	3-11/16	4	4-5/16	4-3/4	5
	B	.70	.72	.73	.74	.74	.82	.89	1.00	1.00

Size of Conduit		Clearance = 1/4"								
		1/2"	3/4'	1"	1-1/4"	1-1/2"	2"	2-1/2"	3"	3-1/2"
1/2"	A	1-3/8	1-1/2	1-5/8	1-13/16	1-15/16	2-1/4	2-9/16	2-7/8	3-1/4
	B	.53	.55	.54	.55	.56	.64	.70	.70	.83
3/4"	A	1-1/2	1-5/8	1-3/4	1-15/16	2-1/16	2-3/8	2-11/16	3	3-3/8
	B	.55	.57	.56	.58	.58	.66	.72	.72	.86
1"	A	1-5/8	1-3/4	1-7/8	2-1/16	2-3/16	2-1/2	2-13/16	3-1/8	3-1/2
	B	.55	.57	.56	.58	.58	.66	.72	.72	.85
1-1/4"	A	1-13/16	1-15/16	2-1/16	2-1/4	2-3/8	2-11/16	3	3-3/8	3-11/16
	B	.55	.58	.58	.58	.59	.67	.73	.79	.86
1-1/2"	A	1-15/16	2-1/16	2-3/16	2-3/8	2-1/2	2-13/16	3-1/8	3-1/2	3-13/16
	B	.56	.58	.58	.59	.59	.68	.74	.79	.87
2"	A	2-1/4	2-3/8	2-1/2	2-11/16	2-13/16	3-1/8	3-7/16	3-3/4	4-1/8
	B	.64	.66	.65	.67	.68	.75	.81	.81	.94
2-1/2"	A	2-9/16	2-11/16	2-13/16	3	3-1/8	3-7/16	3-3/4	4-1/16	4-7/16
	B	.70	.72	.72	.73	.74	.81	.87	.88	1.00
3"	A	2-7/8	3	3-1/8	3-3/8	3-1/2	3-3/4	4-1/16	4-9/16	4-7/8
	B	.70	.72	.72	.79	.79	.81	.88	1.06	1.12
3-1/2"	A	3-1/4	3-3/8	3-1/2	3-11/16	3-13/16	4-1/8	4-7/16	4-7/8	5-1/8
	B	.83	.86	.85	.86	.87	.94	1.00	1.12	1.12

APPROXIMATE SPACING OF CONDUIT BUSHINGS, CHASE NIPPLES AND LOCK NUTS (cont.)

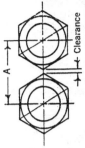

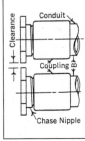

"D" in Simplest Practical Dimension	
Conduit	"D"
1/2	1-1/8
3/4	1-3/8
1	1-5/8
1-1/4	2
1-1/2	2-1/4
2	2-7/8
2-1/2	3-1/2
3	4-5/8
3-1/2	4-7/8

Size of Conduit		Clearance = 3/8"								
		1/2"	3/4'	1"	1-1/4"	1-1/2"	2"	2-1/2"	3"	3-1/2"
1/2"	A	1-1/16	1-5/8	1-3/4	1-15/16	2-1/16	2-3/8	2-11/16	3	3-3/8
	B	.66	.68	.67	.68	.69	.77	.83	.83	.96
3/4"	A	1-5/8	1-3/4	1-7/8	2-1/16	2-3/16	2-1/2	2-13/16	3-1/8	3-1/2
	B	.68	.70	.69	.71	.71	.79	.85	.85	.98
1"	A	1-3/4	1-7/8	2	2-3/16	2-5/16	2-5/8	2-15/16	3-1/4	3-5/8
	B	.68	.69	.69	.71	.71	.78	85	.85	.98
1-1/4"	A	1-15/16	2-1/16	2-3/16	2-3/8	2-1/2	2-13/16	3-1/8	3-1/2	3-13/16
	B	.68	.71	.71	.71	.72	.80	.87	.92	.99
1-1/2"	A	2-1/16	2-3/16	2-5/16	2-1/2	2-5/8	2-15/16	3-1/4	3-5/8	3-15/16
	B	.69	.71	.71	.72	.72	.81	.87	.92	.99
2"	A	2-3/8	2-1/2	2-5/8	2-13/16	2-15/16	3-1/4	3-7/16	3-7/8	4-1/4
	B	.77	.79	.78	.80	.81	.88	.94	1.01	1.07
2-1/2"	A	2-11/16	2-13/16	2-15/16	3-1/8	3-1/4	3-9/16	3-7/8	4-3/16	4-7/16
	B	.83	.85	.85	.86	.87	.94	1.01	1.01	1.13
3"	A	3	3-1/8	3-1/4	3-1/2	3-5/8	3-7/8	4-3/16	4-11/16	5
	B	.83	.85	.85	.92	.92	.94	.99	.99	1.25
3-1/2"	A	3-3/8	3-1/2	3-5/8	3-13/16	3-15/16	4-1/4	4-9/16	5	5-1/4
	B	.96	.98	.98	.99	.99	1.07	1.13	1.25	1.25

Size of Conduit		Clearance = 1/2"								
		1/2"	3/4'	1"	1-1/4"	1-1/2"	2"	2-1/2"	3"	3-1/2"
1/2"	A	1-5/8	1-3/4	1-7/8	2-1/16	2-3/16	2-1/2	2-13/16	3-1/8	3-1/2
	B	.78	.80	.79	.80	.81	.89	.95	.95	1.08
3/4"	A	1-3/4	1-7/8	2	2-3/16	2-5/16	2-5/8	2-15/16	3-1/4	3-5/8
	B	.80	.82	.81	.83	.83	.91	.97	.97	1.10
1"	A	1-7/8	2	2-1/8	2-5/16	2-7/16	2-3/4	3-1/16	3-3/8	3-3/4
	B	.80	.81	.81	.83	.83	.90	97	.97	1.10
1-1/4"	A	2-1/16	2-3/16	2-5/16	2-1/2	2-5/8	2-15/16	3-1/4	3-5/8	3-15/16
	B	.80	.83	.83	.83	.84	.92	.99	1.04	1.11
1-1/2"	A	2-3/16	2-5/16	2-7/16	2-5/8	2-3/4	3-1/16	3-3/8	3-3/4	4-1/16
	B	.81	.83	.83	.84	.84	.93	.98	1.04	1.11
2"	A	2-1/2	2-5/8	2-3/4	2-15/16	3-1/16	3-3/8	3-11/16	4	4-3/8
	B	.89	.91	.90	.92	.93	1.00	1.00	1.06	1.20
2-1/2"	A	2-13/16	2-15/16	3-1/16	3-1/4	3-3/8	3-11/16	4	4-5/16	4-11/16
	B	.95	.97	.97	.98	.99	1.06	1.13	1.13	1.26
3"	A	3-1/8	3-1/4	3-3/8	3-5/8	3-3/4	4	4-5/16	4-13/16	5-1/8
	B	.95	.97	.97	1.04	1.04	1.06	1.14	1.14	1.31
3-1/2"	A	3-1/2	3-5/8	3-3/4	3-15/16	4-1/16	4-1/8	4-11/16	5-1/16	5-3/8
	B	1.01	1.04	1.03	1.04	1.11	1.13	1.31	1.31	1.37

APPROXIMATE SIZES OF CONDUITS, COUPLINGS, CHASE NIPPLES AND BUSHINGS

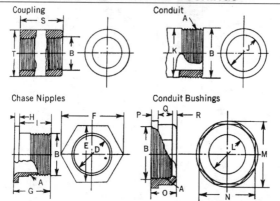

Coupling

Conduit

Chase Nipples

Conduit Bushings

Size of Conduit	A	B	C	D	E	F	G	H	I	J
1/2"	14.0	.82	.85	.62	1.00	1.15	.62	.12	.50	.62
3/4"	14.0	1.02	1.12	.82	1.25	1.44	.81	.19	.62	.82
1"	11.5	1.28	1.67	1.04	1.37	1.59	.94	.25	.69	1.04
1-1/4"	11.5	1.63	2.24	1.38	1.75	2.02	1.06	.25	.81	1.38
1-1/2"	11.5	1.87	2.68	1.61	2.00	2.31	1.12	.31	.81	1.61
2"	11.5	2.34	3.61	2.06	2.50	2.89	1.31	.31	1.00	2.06
2-1/2"	8.0	2.82	5.74	2.46	3.00	3.46	1.44	.37	1.06	2.46
3"	8.0	3.44	7.54	3.06	3.75	4.33	1.50	.37	1.12	3.06
3-1/2"	8.0	3.94	9.00	3.54	4.25	4.91	1.62	.44	1.19	3.54

Size of Conduit	K	L	M	N	O	P	Q	R	S	T
1/2"	.84	.62	1.00	.94	.37	.12	.19	.06	1.37	1.12
3/4"	1.05	.75	1.25	1.12	.44	.12	.25	.06	1.56	1.31
1"	1.31	1.00	1.50	1.37	.50	.16	.25	.09	1.75	1.62
1-1/4"	1.66	1.25	1.81	1.75	.56	.19	.28	.09	2.12	2.00
1-1/2"	1.90	1.50	2.12	2.00	.56	.19	.28	.09	2.50	2.25
2"	2.37	1.94	2.56	2.37	.62	.19	.31	.12	2.62	2.75
2-1/2"	2.87	2.37	3.06	2.87	.75	.25	.37	.12	2.87	3.31
3"	3.50	2.87	3.75	3.50	.81	.25	.37	.19	3.06	3.93
3-1/2"	4.00	3.25	4.25	4.00	1.00	.37	.44	.19	3.62	4.43

DIMENSIONS AND WEIGHTS OF ELECTRICAL METALLIC TUBING

Nominal or Trade Size of Tubing (Inches)	Outside Diameter (Inches)	Minimum Wall Thickness (Inches)	Length (Feet)	Minimum Weight per 100 Feet (Pounds)
3/8	0.577	0.040	10	23
1/2	0.706	0.040	10	28.5
3/4	0.922	0.046	10	43.5
1	1.163	0.054	10	64
1-1/4	1.510	0.061	10	95
1-1/2	1.740	0.061	10	110
2	2.197	0.061	10	140

NOTE: The tolerances are: Length: $\pm$ 1/4-inch
Outside Dia.: $\pm$ .005-inch
Wall Thick.: + 18 percent

DIMENSIONS OF 90-DEGREE ELBOWS

Nominal or Trade Size of Tubing (Inches)	Minimum Radius to Center of Tubing (Inches)	Minimum Straight Length Ls at Each End (Inches)
1/2	4	1-1/2
3/4	4-1/2	1-1/2
1	5-3/4	1-7/8
1-1/4	7-1/4	2
1-1/2	8-1/4	2
2	9-1/2	2

EXPANSION CHARACTERISTICS OF PVC RIGID NON-METALLIC CONDUIT/COEFFICIENT OF THERMAL EXPANSION = 3.38×10^5 in./in./°F

Temperature Change in Degrees F	Length Change in Inches per 100 Ft. of PVC Conduit	Temperature Change in Degrees F	Length Change in Inches per 100 Ft. of PVC Conduit
5	0.20	55	2.23
10	0.41	60	2.43
15	0.61	65	2.64
20	0.81	70	2.84
25	1.01	75	3.04
30	1.22	80	3.24
35	1.42	85	3.45
40	1.62	90	3.65
45	1.83	95	3.85
50	2.03	100	4.06
105	4.26	155	6.29
110	4.46	160	6.49
115	4.66	165	6.69
120	4.87	170	6.90
125	5.07	175	7.10
130	5.27	180	7.30
135	5.48	185	7.50
140	5.68	190	7.71
145	5.88	195	7.91
150	6.08	200	8.11

CHAPTER 4
COMMUNICATIONS

TYPES OF DATA NETWORKS	
10Base2	Is 10MHz Ethernet running over thin, 50 Ohm baseband coaxial cable. 10Base2 is also commonly referred to as thin-Ethernet.
10Base5	Is 10MHz Ethernet running over standard (thick) 50 Ohm baseband coaxial cabling.
10BaseF	Is 10MHz Ethernet running over fiber-optic cabling.
10BaseT	Is 10MHz Ethernet running over unshielded, twisted-pair cabling.
10 Broad36	Is 10MHz Ethernet running through a broadband cable.

STANDARD CONFIGURATIONS FOR SEVERAL COMMON NETWORKS

ATM 155 Mbps	pairs 2 and 4 (pins 1-2, 7-8)
Ethernet 10BaseT	pairs 2 and 3 (pins 1-2, 3-6)
Ethernet 100BaseT4	pairs 2 and 3 (4T+) (pins 1-2, 3-6)
Ethernet 100BaseT8	pairs 1,2,3 and 4 (pins 4-5, 1-2, 3-6, 7-8)
Token-Ring	pairs 1 and 3 (pins 4-5, 3-6)
TP-PMD	pairs 2 and 4 (pins 1-2, 7-8)
100VG-AnyLAN	pairs 1,2,3 and 4 (pins 4-5, 1-2, 3-6, 7-8)

MAXIMUM ATTENUATION OF CATEGORY 3, 4 and 5 CABLING

Maximum Attenuation dB per 1000 ft. @ 20°C

Frequency MHZ	Category 3	Category 4	Category 5
0.064	2.8	2.3	2.2
0.256	4.0	3.4	3.2
0.512	5.6	4.6	4.5
0.772	6.8	5.7	5.5
1.0	7.8	6.5	6.3
4.0	17	13	13
8.0	26	19	18
10.0	30	22	20
16.0	40	27	25
20.0	-	31	28
25.0	-	-	32
31.25	-	-	36
62.5	-	-	52
100	-	-	67

WORST PAIR NEXT LOSS AT SPECIFIC FREQUENCIES

Frequency MHZ	Category 3	Category 4	Category 5
0.150	54	68	74
0.772	43	58	64
1.000	41	56	62
4.000	32	47	53
8.000	28	42	48
10.000	26	41	47
16.000	23	38	44
20.000	-	36	42
25.000	-	-	41
31.250	-	-	40
62.500	-	-	35
100.000	-	-	32

ETHERNET 10BaseT STRAIGHT THRU PATCH CORD

	RJ45 Plug		RJ45 Plug	
T2	1 White/Orange	1	TxData +
R2	2 Orange	2	TxData -
T3	3 White/Green	3	RecvData +
R1	4 Blue	4	
T1	5 White/Blue	5	
R3	6 Green	6	RecvData -
T4	7 White/Brown	7	
R4	8 Brown	8	

ETHERNET 10BaseT CROSSOVER PATCH CORD

This cable is used to cascade hubs, or for connecting two Ethernet stations back-to-back without a hub.

RJ45 Plug	RJ45 Plug
1 Tx+	Rx+ 3
2 Tx -	Rx- 6
3 Rx+	Tx+ 1
6 Rx -	Tx- 2

The cabling administration standard (EIA-606) lists the colors and functions of data cabling as:

Blue	Horizontal voice cables
Brown	Inter-building backbone
Gray	Second-level backbone
Green	Network connections and auxiliary circuits
Orange	Demarcation point, telephone cable from Central Office
Purple	First-level backbone
Red	Key-type telephone systems
Silver or White	Horizontal data cables, computer and PBX equipment
Yellow	Auxiliary, maintenance and security alarms

MINIMUM SEPARATION DISTANCE FROM POWER SOURCE AT 480 V OR LESS

CONDITION	<2kVA	2-5kVA	>5kVA
Unshielded power lines or electrical equipment in proximity to open or non-metal pathways	5 in.	12 in.	24 in.
Unshielded power lines or electrical equipment in proximity to grounded metal conduit pathway	2.5 in.	6 in.	12 in.
Power lines enclosed in a grounded metal conduit (or equivalent shielding) in proximity to grounded metal conduit pathway	-	6 in.	12 in.
Transformers and electric motors Fluorescent lighting	40 in. 12 in.	40 in. 12 in.	40 in. 12 in.

ISDN CONNECTIONS

RJ45 Plug	RJ45 Plug for U+PS2
1 N/C	1 N/C
2 N/C	2 N/C
3 N/C	3 N/C
4 U-loop network connection	4 U-loop network connection
5 U-loop network connection	5 U-loop network connection
6 N/C	6 N/C
7 N/C	7 -48 VDC
8 N/C	8 -48 VDC Return

COMMON TELEPHONE CONNECTIONS

The most common and simplest type of communication installation is the single line telephone. The typical telephone cable (sometimes called quad cable) contains four wires, colored green, red, black, and yellow. A one line telephone requires only two wires to operate. In almost all circumstances, green and red are the two conductors used. In a common four-wire modular connector, the green and red conductors are found in the inside positions, with the black and yellow wires in the outer positions.

As long as the two center conductors of the jack (again, always green and red) are connected to live phone lines, the telephone should operate.

Two-line phones generally use the same four wire cables and jacks. In this case, however, the inside two wires (green and red) carry line 1, and the outside two wires (black and yellow) carry line 2.

COLOR-CODING OF CABLES

The color coding of twisted-pair cable uses a color pattern that identifies not only what conductors make up a pair but also what pair in the sequence it is, relative to other pairs within a multipair sheath. This is also used to determine which conductor in a pair is the *tip* conductor and which is the *ring* conductor. (The tip conductor is the positive conductor, and the ring conductor is the negative conductor.)

The banding scheme uses two opposing colors to represent a single pair. One color is considered the primary while the other color is considered the secondary. For example, given the primary color of white and the secondary color of blue, a single twisted-pair would consist of one cable that is white with blue bands on it. The five primary colors are white, red, black, yellow, and violet.

In multi-pair cables the primary color is responsible for an entire group of pairs (five pairs total). For example, the first five pairs all have the primary color of white. Each of the secondary colors, blue, orange, green, brown, and slate are paired in a banded fashion with white. This continues through the entire primary color scheme for all four primary colors (comprising 25 individual pairs). In larger cables (50 pairs and up), each 25-pair group is wrapped in a pair of ribbons, again representing the groups of primary colors matched with their respective secondary colors. These color coded band markings help cable technicians to quickly identify and properly terminate cable pairs.

EIA COLOR CODE
You should note that the new EIA color code calls for the following color coding:

Pair 1	–	White/Blue (white with blue stripe) and Blue
Pair 2	–	White/Orange and Orange
Pair 3	–	White/Green and Green
Pair 4	–	White/Brown and Brown

TWISTED-PAIR PLUGS AND JACKS

One of the more important factors regarding twisted-pair implementations is the cable jack or cross-connect block. These items are vital slnce without the proper interface, any twisted-pair cable would be relatively useless. In the twisted-pair arena, there are three major types of twisted-pair jacks:

RJ-type connectors (phone plugs)
Pin-connector
Genderless connectors (IBM sexless data connectors)

The RJ-type (registered jack) name generally refers to the standard format used for most telephone jacks. The term pin-connector refers to twisted pair connectors, such as the RS-232 connector, which provide connection through male and female pin receptacles. Genderless connectors are connectors in which there is no separate male or female component; each component can plug into any other similar component.

STANDARD PHONE JACKS

The standard phone jack is specified by a variety of different names, such as RJ and RG, which refer to their physical and electrical characteristics. These jacks consist of a male and a female component. The male component snaps into the female receptacle. The important point to note, however, is the number of conductors each type of jack can support.

Common configurations for phone jacks include support for four, six, or eight conductors. A typical example of a four-conductor jack, supporting two twisted-pairs, would be the one used for connecting most telephone handsets to their receivers.

A common six-conductor jack, supporting three twisted-pairs, is the RJ-11 jack used to connect most telephones to the telephone company or PBX systems. An example of an eight-conductor jack ls the R-45 jack, which is intended for use under the ISDN system as the user-site interface for ISDN terminals.

For building wiring, the six-conductor and eight-conductor jacks are popular, with the eight-conductor jack increasing in popularity, as more corporations install twisted-pair in four-pair bundles for both voice and data. The eight-conductor jack, in addition to being used for ISDN, is also specified by several other popular applications, such as the new IEEE 802.3 10 BaseT standard for Ethernet over twisted-pair.

These types of jacks are often keyed, so that the wrong type of plug cannot be inserted into the jack. There are two kinds of keying — side keying, and shift keying.

Side keying uses a piece of plastic that is extended to one side of the jack. This type is often used when multiple jacks are present.

Shift keying entails shifting the position of the snap connector to the left or right of the jack, rather than leaving it in its usual center position. Shift keying ls more commonly used for data connectors than for voice connectors.

Note that while we say that these jacks are used for certain types of systems (data, voice, etc.) this is not any type of standard. They can be used as you please.

PIN CONNECTORS

There are any number of pin type connectors available. The most familiar type is the RS-232 jack that is commonly used for computer ports. Another popular type of pin connector is the DB type connector, which is the round connector that is commonly used for computer keyboards.

The various types of pin connectors can be used for terminating as few as five (the DB type), or more than 50 (the RS type) conductors.

50-pin *champ* type connectors are often used with twisted-pair cables, when connecting to cross-connect equipment, patch panels, and communications equipment such as is used for networking.

CROSS CONNECTIONS

Cross connections are made at terminal *blocks*. A block is typically a rectangular, white plastic unit, with metal connection points. The most common type is called a punchdown block. This is the kind that you see on the back wall of a business, where the main telephone connections are made. The wire connections are made by pushing the insulated wires into their places. When "punched" down, the connector cuts through the insulation, and makes the appropriate connection.

Connections are made between punch-down blocks by using *patch cords*, which are short lengths of cable that can be terminated into the punch-down slots, or that are equipped with connectors on each end.

When different systems must be connected together, cross-connects are used.

CATEGORY CABLING

Category 1 cable is the old standard type of telephone cable, with four conductors colored green, red, black, and yellow. Also called quad cable.
Category 2 Obsolete.
Category 3 cable is used for digital voice and data transmission rates up to 10 Mbit/s (Megabits per second). Common types of data transmission over this communications cable would be UTP Token Ring (4 Mbit/s) and 10Base-T (10 Mbit/s).
Category 4 Obsolete.
Category 5 cable is used for sending voice and data at speeds up to 100Mbit/s (megabits per second).
Category 6 cable is used for sending data at speeds up to 200 or 250Mbit/s (megabits per second).

INSTALLATION REQUIREMENTS

Article 800 of the NEC covers communication circuits, such as telephone systems and outside wiring for fire and burglar alarm systems. Generally these circuits must be separated from power circuits and grounded. In addition, all such circuits that run out of doors (even if only partially) must be provided with circuit protectors (surge or voltage supressors).

The requirements for these installations are as follows:

CONDUCTORS ENTERING BUILDINGS

If communications and power conductors are supported by the same pole, or run parallel in span, the following conditions must be met:

1. Wherever possible, communications conductors should be located below power conductors.
2. Communications conductors cannot be connected to crossarms.
3. Power service drops must be separated from communications service drops by at least 12 inches.

Above roofs, communications conductors must have the following clearances:

1. Flat roofs: 8 feet.
2. Garages and other auxiliary buildings: None required.
3. Overhangs, where no more than 4 feet of communications cable will run over the area: 18 inches.
4. Where the roof slope is 4 inches rise for every 12 inches horizontally: 3 feet.

Underground communications conductors must be separated from power conductors in manhole or handholes by brick, concrete, or tile partitions.

Communications conductors should be kept at least 6 feet away from lightning protection system conductors.

CIRCUIT PROTECTION

Protectors are surge arresters designed for the specific requirements of communications circuits. They are required for all aerial circuits not confined with a *block*. (Block here means city block.) They must be installed on all circuits with a block that could accidentally contact power circuits over 300 volts to ground. They must also be listed for the type of installation.

Other requirements are the following:

Metal sheaths of any communications cables must be grounded or interrupted with an insulating joint as close as practicable to the point where they enter any building (such point of entrance being the place where the communications cable emerges through an exterior wall or concrete floor slab, or from a grounded rigid or intermediate metal conduit).

Grounding conductors for communications circuits must be copper or some other corrosion-resistant material, and have insulation suitable for the area in which it is installed.

Communications grounding conductors may be no smaller than #14 (AWG).

The grounding conductor must be run as directly as possible to the grounding electrode, and be protected if necessary.

If the grounding conductor is protected by metal raceway, it must be bonded to the grounding conductor on both ends.

CIRCUIT PROTECTION (cont.)

Grounding electrodes for communications ground may be any of the following:
1. The grounding electrode of an electrical power system.
2. A grounded interior metal piping system (Avoid gas piping systems for obvious reasons.)
3. Metal power service raceway.
4. Power service equipment enclosures.
5. A separate grounding electrode.

If the building being served has no grounding electrode system, the following can be used as a grounding electrode:
1. Any acceptable power system grounding electrode. (See Section 250-81.)
2. A grounded metal structure.
3. A ground rod or pipe at least 5 feet long and 1/2 inch in diameter. This rod should be driven into damp (if possible) earth, and kept separate from any lightning protection system grounds or conductors.

Connections to grounding electrodes must be made with approved means.

If the power and communications systems use separate grounding electrodes, they must be bonded together with a No. 6 copper conductor. Other electrodes may be bonded also. This is not required for mobile homes.

For mobile homes, if there is no service equipment or disconnect within 30 feet of the mobile home wall, the communications circuit must have its own grounding electrode. In this case, or if the mobile home is connected with cord and plug, the communications circuit protector must be bonded to the mobile home frame or grounding terminal with a copper conductor no smaller than No. 12.

INTERIOR COMMUNICATIONS CONDUCTORS

Communications conductors must be kept at least 2 inches away from power or Class 1 conductors, unless they are permanently separated from them or unless the power or Class 1 conductors are enclosed in one of the following:
1. Raceway.
2. Type AC, MC, UF, NM, or NM cable, or metal-sheathed cable.

Communications cables are allowed in the same raceway, box, or cable with any of the following:
1. Class 2 and 3 remote-control, signaling, and power-limited circuits.
2. Power-limited fire protective signaling systems.
3. Conductive or nonconductive optical fiber cables.
4. Community antenna television and radio distribution systems.

Communications conductors are not allowed to be in the same raceway or fitting with power or Class 1 circuits.

Communications conductors are not allowed to be supported by raceways unless the raceway runs directly to the piece of equipment the communications circuit serves.

Openings through fire-resistant floors, walls, etc. must be sealed with an appropriate firestopping material.

Any communications cables used in plenums or environmental air-handling spaces must be listed for such use.

STANDARD TELECOM COLOR CODING

PAIR #	TIP (+) COLOR	RING (–) COLOR
1	White	Blue
2	White	Orange
3	White	Green
4	White	Brown
5	White	Slate
6	Red	Blue
7	Red	Orange
8	Red	Green
9	Red	Brown
10	Red	Slate
11	Black	Blue
12	Black	Orange
13	Black	Green
14	Black	Brown
15	Black	Slate
16	Yellow	Blue
17	Yellow	Orange
18	Yellow	Green
19	Yellow	Brown
20	Yellow	Slate
21	Violet	Blue
22	Violet	Orange
23	Violet	Green
24	Violet	Brown
25	Violet	Slate

25-PAIR COLOR CODING/ISDN CONTACT ASSIGNMENTS

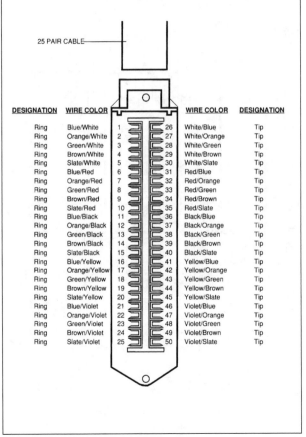

25 PAIR CABLE

DESIGNATION	WIRE COLOR			WIRE COLOR	DESIGNATION
Ring	Blue/White	1	26	White/Blue	Tip
Ring	Orange/White	2	27	White/Orange	Tip
Ring	Green/White	3	28	White/Green	Tip
Ring	Brown/White	4	29	White/Brown	Tip
Ring	Slate/White	5	30	White/Slate	Tip
Ring	Blue/Red	6	31	Red/Blue	Tip
Ring	Orange/Red	7	32	Red/Orange	Tip
Ring	Green/Red	8	33	Red/Green	Tip
Ring	Brown/Red	9	34	Red/Brown	Tip
Ring	Slate/Red	10	35	Red/Slate	Tip
Ring	Blue/Black	11	36	Black/Blue	Tip
Ring	Orange/Black	12	37	Black/Orange	Tip
Ring	Green/Black	13	38	Black/Green	Tip
Ring	Brown/Black	14	39	Black/Brown	Tip
Ring	Slate/Black	15	40	Black/Slate	Tip
Ring	Blue/Yellow	16	41	Yellow/Blue	Tip
Ring	Orange/Yellow	17	42	Yellow/Orange	Tip
Ring	Green/Yellow	18	43	Yellow/Green	Tip
Ring	Brown/Yellow	19	44	Yellow/Brown	Tip
Ring	Slate/Yellow	20	45	Yellow/Slate	Tip
Ring	Blue/Violet	21	46	Violet/Blue	Tip
Ring	Orange/Violet	22	47	Violet/Orange	Tip
Ring	Green/Violet	23	48	Violet/Green	Tip
Ring	Brown/Violet	24	49	Violet/Brown	Tip
Ring	Slate/Violet	25	50	Violet/Slate	Tip

66 BLOCK WIRING AND CABLE COLOR CODING

PAIR CODE	SIDE #1			SIDE #2
Pair 1	Tip 26	White/Blue		White/Blue
	Ring 1	Blue/White		Blue/White
Pair 2	Tip 27	White/Orange		White/Orange
	Ring 2	Orange/White		Orange/White
Pair 3	Tip 28	White/Green		White/Green
	Ring 3	Green/White		Green/White
Pair 4	Tip 29	White/Brown		White/Brown
	Ring 4	Brown/White		Brown/White
Pair 5	Tip 30	White/Slate		White/Slate
	Ring 5	Slate/White		Slate/White
Pair 6	Tip 31	Red/Blue		Red/Blue
	Ring 6	Blue/Red		Blue/Red
Pair 7	Tip 32	Red/Orange		Red/Orange
	Ring 7	Orange/Red		Orange/Red
Pair 8	Tip 33	Red/Green		Red/Green
	Ring 8	Green/Red		Green/Red
Pair 9	Tip 34	Red/Brown		Red/Brown
	Ring 9	Brown/Red		Brown/Red
Pair 10	Tip 35	Red/Slate		Red/Slate
	Ring 10	Slate/Red		Slate/Red
Pair 11	Tip 36	Black/Blue		Black/Blue
	Ring 11	Blue/Black		Blue/Black
Pair 12	Tip 37	Black/Orange		Black/Orange
	Ring 12	Orange/Black		Orange/Black

TOP

Top color labels (left to right):
Black/Green, Green/Black, Black/Brown, Brown/Black, Black/Slate, Slate/Black, Yellow/Blue, Blue/Yellow, Yellow/Orange, Orange/Yellow, Yellow/Green, Green/Yellow, Yellow/Brown, Brown/Yellow, Yellow/Slate, Slate/Yellow, Violet/Blue, Blue/Violet, Violet/Orange, Orange/Violet, Violet/Green, Green/Violet, Violet/Brown, Brown/Violet, Violet/Slate, Slate/Violet

Pair 13	Tip 38	Black/Green
	Ring 13	Green/Black
Pair 14	Tip 39	Black/Brown
	Ring 14	Brown/Black
Pair 15	Tip 40	Black/Slate
	Ring 15	Slate/Black
Pair 16	Tip 41	Yellow/Blue
	Ring 16	Blue/Yellow
Pair 17	Tip 42	Yellow/Orange
	Ring 17	Orange/Yellow
Pair 18	Tip 43	Yellow/Green
	Ring 18	Green/Yellow
Pair 19	Tip 44	Yellow/Brown
	Ring 19	Brown/Yellow
Pair 20	Tip 45	Yellow/Slate
	Ring 20	Slate/Yellow
Pair 21	Tip 46	Violet/Blue
	Ring 21	Blue/Violet
Pair 22	Tip 47	Violet/Orange
	Ring 22	Orange/Violet
Pair 23	Tip 48	Violet/Green
	Ring 23	Green/Violet
Pair 24	Tip 49	Violet/Brown
	Ring 24	Brown/Violet
Pair 25	Tip 50	Violet/Slate
	Ring 25	Slate/Violet

ISDN ASSIGNMENT OF CONTACT NUMBERS

Table-Contact assignments for plugs and jacks:

Contact Number	TE	NT	Polarity
1	Power source 3	Power sink 3	+
2	Power source 3	Power sink 3	-
3	Transmit	Receive	+
4	Receive	Transmit	+
5	Receive	Transmit	-
6	Transmit	Receive	-
7	Power sink 2	Power source 2	-
8	Power sink 2	Power source 2	+

8P8C (IISDN)

PS3 T-R PS2
/\ /RT\ /\

- - + + - - + +
| | | | | | | |
1 2 3 4 5 6 7 8

TYPICAL WIRING METHODS

LOOP SERIES WIRING

Protector | Service Entry | Jack | Jack | Jack

PARALLEL DISTRIBUTION WIRING

Service Entry | Jack | Jack | Jack | Protector

2-LINE SYSTEM

Protector | Service Entry | (2-Line Modular) | 2-Line Telephone

ELECTRONIC KEY SYSTEMS

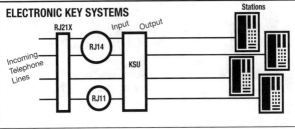

RJ21X | Input | Output | Stations | RJ14 | KSU | Incoming Telephone Lines | RJ11

MODULAR JACK STYLES

8-Position

8-Position Keyed

6-Position

6-Position Modified

There are four basic modular jack styles. The 8-position and 8-position keyed modular jacks are commonly and incorrectly referred to as RJ45 and keyed RJ45 (respectively). The 6-position modular jack is commonly referred to as RJ11. Using these terms can sometimes lead to confusion since the RJ designations actually refer to very specific wiring configurations called Universal Service Ordering Codes (USOC). The designation 'RJ' means Registered Jack. Each of these 3 basic jack styles can be wired for different RJ configurations. For example, the 6-position jack can be wired as a RJ11C (1-Pair), RJ14C (2-Pair), or RJ25C (3-Pair) configuration. An 8-position jack can be wired for configurations such as RJ61C (4-Pair) and RJ48C. The keyed 8-position jack can be wired for RJ45S, RJ46S and RJ47S. The fourth modular jack style is a modified version of the 6-position jack (modified modular jack or MMJ). It was designed by DEC along with the modified modular plug (MMP) to eliminate the possibility of connecting DEC data equipment to voice lines and vice versa.

COMMON WIRING CONFIGURATIONS

The TIA and AT&T wiring schemes are the two that have been adopted by EIA/TIA-568. They are nearly identical except that pairs two and three are reversed. TIA is the preferred scheme because it is compatible with 1- or 2-pair USOC Systems. Either configuration can be used for Integrated Services Digital Network (ISDN) applications.

Pair ID	PIN #
T1	5
R1	4
T2	3
R2	6
T3	1
R3	2
T4	7
R4	8

TIA (T568A)

Pair ID	PIN #
T1	5
R1	4
T2	1
R2	2
T3	3
R3	6
T4	7
R4	8

AT&T (T568B)

COMMON WIRING CONFIGURATIONS (cont.)

USOC wiring is available for 1-, 2-, 3-, or 4-pair systems. Pair 1 occupies the center conductors, pair 2 occupies the next two contacts out, etc. One advantage to this scheme is that a 6-position plug configured with 1, 2, or 3 pairs can be inserted into an 8-position jack and maintain pair continuity; a note of warning though, pins 1 and 8 on the jack may become damaged from this practice. A disadvantage is the poor transmission performance associated with this type of pair sequence.

Pair ID	PIN #
T1	5
R1	4
T2	3
R2	6
T3	2
R3	7
T4	1
R4	8

USOC 4-Pair

Pair ID	PIN #
T1	4
R1	3
T2	2
R2	5
T3	1
R3	6

USOC 1-, 2-, or 3-Pair

ETHERNET 10BaseT

Ethernet 10BaseT wiring specifies an 8-position jack but uses only two pairs. These are pairs two and three of TIA schemes.

Pair ID	PIN #
T1	1
R1	2
T2	3
R2	6

COMMON WIRING CONFIGURATIONS (cont.)

IBM Token-Ring

IBM Token-Ring wiring uses either an 8-position or 6-position jack. The 8-position format is compatible with TIA wiring schemes. The 6-position is compatible with 2 or 3-pair USOC wiring.

Pair ID	PIN #
T1	5
R1	4
T2	3
R2	6

DEC 3-Pair

DEC custom-designed wiring scheme is unique.

Pair ID	PIN #
T1	2
R1	3
T2	5
R2	4
T3	1
R3	6

MODULAR PLUG PAIR CONFIGURATIONS

It is important that the pairing of wires in the modular plug match the pairs in the modular jack as well as the horizontal and backbone wiring. If they don't, the data being transmitted may be paired with incompatible signals.

STRAIGHT THROUGH OR REVERSED?

Modular cords are used for two basic applications. One application uses them for patching between modular patch panels. When used in this manner, modular cords should always be wired "straight-through" (pin 1 to pin 1, pin 2 to pin 2, pin 3 to pin 3, etc.). The second major application uses modular cords to connect the workstation equipment (PC, phone, FAX, etc.) to the modular outlet. These modular cords may either be wired "straight-through" or "reversed" (pin 1 to pin 6, pin 2 to pin 5, pin 3 to pin 4, etc.) depending on the system manufacturer's specifications. This "reversed" wiring is typical for most voice systems. The following is a guide to determine what type of modular cord you have.

HOW TO READ A MODULAR CORD

Align the plugs side-by-side with the contacts facing you and compare the wire colors from left to right. If the colors appear in the same order on both plugs, the cord is wired "straight-through" If the colors appear reversed on the second plug (from right to left) the cord is wired "reversed."

CHAPTER 5
FIBER OPTICS

TYPICAL OPTICAL BUDGETS

Wavelength [nm]	Type of Source	Fiber Core Diameter [μm]	Typical Optical Power Budget, [dB]	Spectral Width [nm]
850	LED	50	12	
	LED	6.25	16	
	LED	100	21	
	LED	any		30-50
	laser	all	30	
1300	LED	50	20	
	LED	62.5	24	
	LED	100	28	
	LED	any		60-190
1300	laser [multimode]	all	50	
1300	laser [singlemode]	9	27	0.5 - 5.0

MAXIMUM VERTICAL RISE DISTANCES

Application	Feet
1 fiber in raceway or tray	90
2 fiber in duct or conduit	50-90
Multifiber (6-12) cables	50-375
Heavy duty cables	1000-1640

FIBER OPTIC CONNECTORS
DATA COMMUNICATION STYLES

Style	Contact?	Keyed?	Pull Proof?	Wiggle-Proof?	Loss	Cost
ST®	Y	Y	N	N	0.3	6-10
906 SMA	N	N	Y	N	1.0	6-10
FDDI	Y	Y	Y	N	1.0	12-19
mini-BNC	N	N	N	N	—	10
905 SMA	N	N	Y	N	1.0	6-10
biconic	N	N	Y	N	1.0	8-25
ESCON	Y	Y	N	N	1.0	25
SC	Y	Y	Y	Y	0.3	7-14

TELEPHONE AND HIGH PERFORMANCE STYLES

Style	Contact?	Keyed?	Pull Proof?	Wiggle-Proof?	Loss	Cost
ST®	Y	Y	N	N	0.3	6-10
biconic	N	N	Y	N	1.0	8-25
keyed biconic	N	Y	Y	N	1.0	8-25
FC/PC	Y	Y	Y	Y	0.3	4-15
FC	Y	Y	Y	Y	0.3	4-15
D4	Y	Y	Y	Y	0.3	12
SC	Y	Y	Y	Y	0.3	7-14

TYPICAL BANDWIDTH — DISTANCE PRODUCTS

Type of Fiber	Wavelength	Bandwidth - Distance Product
multimode step index POF	660 nm	5 MHz-km
multimode step index glass	850 nm	20 MHz-km
multimode graded index glass	850 nm	600 MHz-km
multimode graded index glass	1300 nm	1000-2500 MHz-km
singlemode glass	1310 nm	76,800-300,000 Mbps-km

OPTICAL CABLE JACKET MATERIALS AND THEIR PROPERTIES

Jacket Materials	Properties
PVC	Affords normal mechanical protection. Usually specified for indoor use and general-purpose applications.
Hypalon	Has most of neoprene's properties, including ability to withstand extreme environments and flame retardancy. Has better thermal stability, and even greater oxidation and ozone resistance. Hypalon has superior resistance to radiation.
Polyethylene	Used in telephone cables. A tough, chemical- and moisture-resistant, relatively low-cost material. Since it burns, it is infrequently used in electronic applications.
Polyurethane	Has excellent abrasion resistance and low-temperature flexibility.
Thermoplastic Elastomer (TPE)	A less expensive jacketing material than neoprene or hypalon. Has many of the characteristics of rubber, along with excellent mechanical and chemical properties.
Nylon	Generally used over single conductors to improve their physical properties.

OPTICAL CABLE JACKET MATERIALS/PROPERTIES (cont.)

Jacket Materials	Properties
Kynar	A tough, abrasion- and cut- through-resistant, thermally (polyvinylidene fluoride)stable and self-extinguishing material. It has low-smoke emission and is resistant to most chemicals. Its inherent stiffness limits its use as a jacket material. It has been approved for low-smoke applications.
Teflon FEP	Specified in fire alarm signal system cables. It will not emit smoke even when exposed to direct flame, is suitable for use at continuous temperatures of 200˚C and is chemically inert.
Tefzel	Like Teflon FEP, it is a fluorocarbon and has many of its properties. Rated for 150˚C, it is a tough, self-extinguishing material.
Irradiated Cross-Linked	Rated for 150˚C operation. Cross-linking changes ther-moplastic polyethylene to a thermosetting material with greater resistance to environmental stress cracking, cut-through, ozone, solvents and soldering than either low- or high-density polyethylene.
Zero Halogen Thermoplastic	A thermoplastic material with excellent flame retardancy properties. Does not emit toxic fumes when it burns. Originally designed for shipboard fiber applications, it can be used for any enclosed environment.

NEC CLASSIFICATIONS FOR OPTICAL CABLES

Application	Cable marking	UL test	Jacket type
General Purpose	OFN	UL-1581	PVC or
	OFC	UL-1581	Zero Halogen
Riser	OFNR	UL-1666	PVC or
	OFCR	UL-1666	Zero Halogen
Plenum	OFNP	UL-910	RVC or
	OFCP	UL-910	Fluorocarbons

Cable Marking Explanation

OFN	Optical fiber, nonconductive (all dielectric)
OFC	Optical fiber, conductive (metal strength members)
OFNR	Optical fiber, nonconductive, riser
OFCR	Optical fiber, conductive, riser
OFNP	Optical fiber, nonconductive, plenum
OFCP	Optical fiber, conductive, plenum

COMMON FIBER TYPES

Item Number	Diameter Core	Diameter Cladding	Index Profile	Primary Buffer Diameter	Attenuation (dB/km)	Bandwidth (MHz/km)	Numerical Aperture
1)	50μm	125μm	Graded	250μm	•3 @ 0.85μm •1 @ 1.3μm	200-800 @ 0.85μm 200-800 @ 1.3μm	0.20
2)	50μm	125μm	Graded	500μm	•4 @ 0.85μm •2 @ 1.3μm	200-800 @ 0.85μm 200-800 @ 1.3μm	0.20
3)	62.5μm	125μm	Graded	250μm	•3.5 @ 0.85μm •1.5 @ 1.3μm	100-300 @ 0.85μm 100-800 @ 1.3μm	0.275
4)	62.5μm	125μm	Graded	500μm	•3.5 @ 0.85μm •1.5 @ 1.3μm	100-300 @ 0.85μm 100-800 @ 1.3μm	0.275
5)	85μm	125μm	Graded	500μm	•4 @ 0.85μm •2 @ 1.3μm	100-200 @ 0.85μm 200-400 @ 1.3μm	0.26
6)	100μm	140μm	Graded	500μm	•5 @ 0.85μm •3 @ 1.3μm	100-300 @ 0.85μm 100-500 @ 1.3μm	0.30
7)	100μm	140μm	Step	500μm	•10 @ 0.85μm	20 @ 0.85μm	0.24
8)	100μm	140μm	Graded	160μm	•6 @ 0.85μm	100 @ 0.85μm	0.30
9)	200μm	230μm	Step(HCS)	500μm	•8 @ 0.85μm	17 @ 0.85μm	0.37
10)	200μm	240μm	Step	500μm	•10 @ 0.85μm	20 @ 0.85μm	0.24
11)	200μm	380μm	Step(PCS)	600μm	•10 @ 0.85μm	8 @ 0.85μm	0.40

Note items 7 and 9 are radiation-hard fibers. Item 8 is a high-temperature fiber.

COMPARISON OF LASER AND LED LIGHT SOURCES

Characteristic	LED	Laser
Output power	Lower	Higher
Speed	Slower	Faster
Output pattern (NA)	Higher	Lower
Spectral width	Wider	Narrower
Single-mode compatibility	No	Yes
Ease of Use	Easier	Harder
Cost	Lower	Higher

COMPARISON OF BUFFER TYPES

Cable Parameter	Cable Structure	
	Loose Tube	Tight Buffer
Bend Radius	Larger	Smaller
Diameter	Larger	Smaller
Tensile Strength, Installation	Higher	Lower
Impact Resistance	Lower	Higher
Crush Resistance	Lower	Higher
Attenuation Change at Low Temperatures	Lower	Higher

INDICES OF REFRACTION

Material	Index	Light Velocity(km/s)
Vacuum	1.0	300,000
Air	1.0003	300,000
Water	1.33	225,000
Fused quartz	1.46	205,000
Glass	1.5	200,000
Diamond	2.0	150,000
Silicon	3.4	88,000
Gallium arsenide	3.6	83,000

TRANSMISSION RATES — DIGITAL TELEPHONE

Medium	Bit Rate (Mbps)	Voice Channels	Repeater Spacing (km)
Coaxial	1.5	24	1-2
	3.1	48	
	6.3	96	
	45	672	
	90	1,344	
Fiber	45	672	6-15 (multimode)
	90	1,344	30-40+ (single mode)
	180	2,688	
	405 to 435	6,048	
	565	8,064	
	1,700	24,192	

MISMATCHED FIBER CONNECTION LOSSES (Excess loss in dB)

Receiving Fiber	Transmitting Fiber		
	62.5/125	85-125	100/140
50-125	0.9-1.6	3.0-4.6	4.7-9
62.5/125	—	0.9	2.1-4.1
85/125	—	—	0.9-1.4

POWER LEVELS OF FIBER OPTIC COMMUNICATION SYSTEMS

Network Type	Wavelength (nm)	Power Range (dBm)	Power Range (W)
Telecom	1300, 1550	+3 to -45	50nW to 2mW
Datacom	665, 790, 850, 1300	-10 to -30	1uW to 100uW
CATV	1300, 1550	+10 to -6	250uW to 10mW

DETECTORS USED IN FIBER OPTIC POWER METERS

Detector Type	Wavelength Range (nm)	Power Range (dBm)	Comments
Silicon	400-1100	+10 to -70	
Germanium	800-1600	+10 to -60	-70 with small area detectors, +30 with attenuator windows
InGaAs	800-1600	+10 to -70	Small area detectors may overload at high power (>.0 dBm)

FIBER OPTIC TESTING REQUIREMENTS

Test Parameter	Instrument
Optical power (source output, receiver signal level)	Fiber optic power meter
Attenuation or loss of fibers, cables, and connectors	FO power meter and source, test kit or OLTS (optical loss test set)
Source wavelength*	FO spectrum analyzer
Backscatter (loss, length, fault location)	Optical time domain reflectometer (OTDR)
Fault location	OTDR, visual cable fault locator
Bandwidth/dispersion* (modal and chromatic)	Bandwidth tester or simulation software

* Rarely tested in the field

TYPICAL CABLE SYSTEM FAULTS

Fault	Cause	Equipment	Remedy
Bad connector	Dirt or damage	Microscope	Cleaning/polishing retermination
Bad pigtail	Pigtail kinked	Visual fault locator	Straighten kink
Localized cable attenuation	Kinked cable	OTDR	Straighten kink
Distributed increase in cable attenuation	Defective cable or installation specifications exceeded	OTDR	Reduce stress/replace
Lossy splice	Increase in splice Loss due to fiber stress in closure	OTDR Visual fault locator	Open and redress
Fiber break	Cable damage	OTDR Visual fault locator	Repair/replace

OPTICAL CABLE CRUSH STRENGTHS

Characteristic	Type of Cable	Pounds/Inch
Long-term crush load	>6 fibers/cable	57-400
	1-2 fiber cables	314-400
	Armored cables	450
Short-term crush load	>6 fibers/cable	343-900
	1-2 fiber cables	300-800
	Armored cables	600

MOST COMMON CAUSES OF FAILURES IN FIBER OPTIC LANs

1. Broken fibers at connector joints
2. Broken fibers at patch panels
3. Cables damaged at patch panels
4. Fibers broken at patch panels
5. Cables cut in ceilings and walls
6. Cables cut through outside construction
7. Contaminated connections
8. Broken jumpers
9. Too much loss
10. Too little loss (overdriving the receiver)
11. Improper cable rolls
12. Miskeyed connectors
13. Transmission equipment failure
14. Power failure

FIBER OPTIC LABOR UNITS

Labor Item	Labor Units (Hours) Normal	Difficult
Optical fiber cables, per foot:		
1-4 fibers, in conduit	0.016	0.02
1-4 fibers, accessible locations	0.014	0.018
12-24 fibers, in conduit	0.02	0.025
12-24 fibers, accessible locations	0.018	0.023
48 fibers, in conduit	0.03	0.038
48 fibers, accessible locations	0.025	0.031
72 fibers, in conduit	0.04	0.05
72 fibers, accessible locations	0.032	0.04
144 fibers, in conduit	0.05	0.065
144 fibers, accessible locations	0.04	0.05
Hybrid cables:		
1-4 fibers, in conduit	0.02	0.025
1-4 fibers, accessible locations	0.017	0.021
12-24 fibers, in conduit	0.024	0.03
12-24 fibers, accessible locations	0.022	0.028
Testing, per fiber	0.10	0.20
Splices, including prep and failures, trained workers:		
fusion	0.20	0.30
mechanical	0.30	0.40
array splice, 12 fibers	0.80	1.20
Coupler (connector-connector)	0.15	0.25
Terminations, including prep and failures, trained workers:		
polishing required	0.35	0.55
no-polish connectors	0.25	0.40
FDDI dual connector, including terminations	0.70	0.95
Miscellaneous:		
cross-connect box, 144 fibers, not including splices	3.00	4.00
splice cabinet	2.00	2.50
splice case	1.80	2.25
breakout kit, 6 fibers	1.00	1.40
tie-wraps	0.01	0.02
wire markers	0.01	0.01

FIBER OPTIC SAFETY RULES

1. Keep all food and beverages out of the work area. If fiber particles are ingested, they can cause internal hemorrhaging.
2. Wear disposable aprons to minimize fiber particles on your clothing. Fiber particles on your clothing can later get into food, drinks, and/or be ingested by other means.
3. Always wear protective gloves and safety glasses with side shields. Treat fiber optic splinters the same as you would glass splinters.
4. Never look directly into the end of fiber cables until you are positive that there is no light source at the other end. Use a fiber optic power meter to make certain the fiber is dark. When using an optical tracer or continuity checker, look at the fiber from an angle at least six inches away from your eye to determine if the visible light is present.
5. Only work in well ventilated areas.
6. Contact wearers must not handle their lenses until they have thoroughly washed their hands.
7. Do not touch your eyes while working with fiber optic systems until your hands have been thoroughly washed.

CABLE SELECTION CRITERIA

1. Current and future bandwidth requirements
2. Acceptable attenuation rate
3. Length of cable
4. Cost of installation
5. Mechanical requirements (ruggedness, flexibility, flame retardance, low smoke, cut-through resistance)
6. UL/NEC requirements
7. Signal source (coupling efficiency, power output, receiver sensitivity)
8. Connectors and terminations
9. Cable dimension requirements
10. Physical environment (temperature, moisture, location)
11. Compatibility with any existing systems

MAXIMUM RECOMMENDED INSTALLATION LOADS

Application	Pounds Force
1 fiber in raceway or tray	67
1 fiber in duct or conduit	125
2 fiber in duct or conduit	200
Multifiber (6-12) cables	250-500
Direct burial cables	600-800
Lashed aerial cables	>300
Self-support aerial cables	>600

FIBER OPTIC DATA NETWORK STANDARDS

Network	IEEE802.3 FOIRL	IEEE802.3 10baseF	IEEE802.5 Token Ring	ANSI X3T9.5 FDDI	ESCON IBM
Bitrate (MB/s)	10	10	4/16	100	200
Architecture	Link	Star	Ring	Ring	Branch
Fiber type	MM, 62.5	MM, 62.5	MM, 62.5	MM/SM	MM/SM
Link length (km)	2	—	—	2/60	3/20
Wavelength (nm)	850	850	850	1300	1300
Margin (dB, MM/SM)	8	—	12	11/27	8*(11)/16
Fiber FW (mHz-km)	150	150	150	500	500
Connector	SMA	ST	FDDI	FDDI	ESCON

FIBER TYPES AND SPECIFICATIONS

Fiber Type Bandwidth	Core/Cladding	Attenuation Coefficient (dBkm)			
	Diameter (m)	850 nm	1300 nm	1550 nm	(MHz-km)
Multimode/Plastic	1mm	(1 dB/m @665 nm)			Low
Multimode/Step Index	200/240	6			50
Multimode/Graded Index	50/125	3	1		600
	62.5/125	3	1		500
	85/125	3	1		500
	100/140	3	1		300
Single mode	8-9/125		0.5	0.3	high

COMMON FIBER OPTIC CONNECTORS

SFR Type

SMA905

SMA906

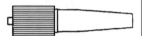

D4 CERAMIC CONNECTOR

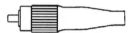

FC Cylindrical with metal coupling

BICONIC

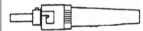

ST Cylindrical with twist lock coupling

SC Square, keyed connector

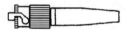

MINI BNC Cylindrical with twist lock coupling

FDDI
Duplex connector with fixed shroud, keyed

ESCON
Duplex connector with retractable shroud

CHAPTER 6
MOTORS

DESIGNING MOTOR CIRCUITS

For one motor:
1. Determine full-load current of motor(s).
2. Multiply full-load current x 1.25 to determine minimum conductor ampacity.
3. Determine wire size.
4. Determine conduit size.
5. Determine minimum fuse or circuit breaker size.
6. Determine overload rating.

For more than one motor:
1. Perform steps 1 through 6 as shown above for each motor.
2. Add full-load current of all motors, plus 25% of the full-load current of the largest motor to determine minimum conductor ampacity.
3. Determine wire size.
4. Determine conduit size.
5. Add the fuse or circuit breaker size of the largest motor, plus the full-load currents of all other motors to determine the maximum fuse or circuit breaker size for the feeder.

MOTOR CONTROL CIRCUITS

Magnetic starter with one stop-start station and a pilot lamp which burns to indicate that the motor is running.

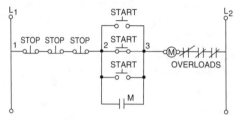

Magnetic starter with three stop-start stations.

Diagrammatic representation of a magnetic three-phase starter with one start-stop station.

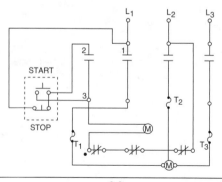

MOTOR CONTROL CIRCUITS (cont.)

Jogging using a selector push button.

Magnetic starter with a plugging switch.

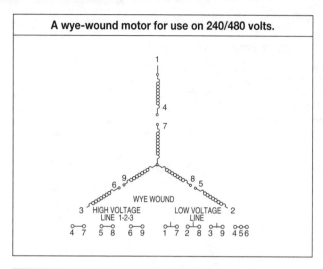

A wye-wound motor for use on 240/480 volts.

WYE WOUND

HIGH VOLTAGE
LINE 1-2-3

4 7   5 8   6 9

LOW VOLTAGE
LINE

1 7   2 8   3 9   4 5 6

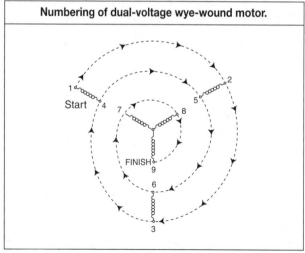

Numbering of dual-voltage wye-wound motor.

Start

FINISH

TIPS ON SELECTING MOTORS

First step in selecting a motor for a particular drive is to obtain data listed below. Motors generally operate at best power factor and efficiency when fully loaded.

TORQUE: Starting torque needed by load must be less than required starting torque of proposed motor. Motor torque must never fall below driven machine's torque needs in going from standstill to full speed.

Torque requirements of some loads may fluctuate between wide limits. Although average torque may be low, many torque peaks may be well above full-load torque. If load torque impulses are repeated frequently (air compressor), it's best to use a high-slip motor with a flywheel. But if load is generally steady at full load, you can use a more efficient low-slip motor. Only in this case any intermittent load peaks are taken directly by the motor and reflect back into power system. Also breakdown (maximum motor torque) must be higher than load-peak torque.

ENCLOSURES: Atmospheric conditions surrounding motor determine enclosure used. The more enclosed a motor, the more it costs and the hotter it tends to run. Totally-enclosed motors may require a larger frame size for a given hp than open or protected motors.

INSULATION: This, likewise, is determined by surrounding atmosphere and operating temperature. Ambient (room) temperature is generally assumed to be 40°C. Total temperature motor reaches directly influences insulation life. Each 10°C rise in Max. temp halves effective life of *Class A* and *B* insulations.

Motor temperature rise is maximum temperature (over ambient) measured with an external thermometer. "Hot-spot" allowance takes care of temperature difference between external reading and hottest spot within windings. Service-factor allows for continuous overload - 15% for general-purpose open, protected or drip-proof motors.

VARIABLE CYCLE: Where load varies according to some regular cycle, it would not be economical to select a motor that matches the peak load.

Instead (for AC induction motors where speed is not varied) calculate hp needed on the *root-means-square* (rms) basis, example right. *Rms* hp is equivalent continous hp that would produce same heat in motor as cycle operation. Torque-speed relation of motor should still match that of load.

FACTS TO CONSIDER

REQUIREMENTS OF DRIVEN MACHINE:
1. Hp needed
2. Torque range
3. Operating cycle-frequency of starts and stops
4. Speed
5. Operating position-horizontal, vertical or tilted
6. Direction of rotation
7. Endplay and thrust
8. Ambient (room) temperature
9. Surrounding conditions water, gas, corrosion, dust, outdoor, etc.

ELECTRICAL SUPPLY:
1. Voltage of power system
2. Number of phases
3. Frequency
4. Limitations on starting current
5. Effect of demand, energy on power rates

SQUIRREL-CAGE MOTORS

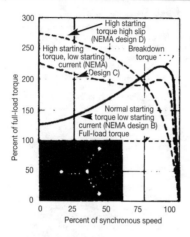

Squirrel-cage induction motors are classed by National Electrical Manufacturers Assn. (NEMA) according to locked-rotor torque, breakdown torque, slip, starting current, etc. Common types are Class B, C and D.

CLASS B is most common type, has normal starting torque, low starting current. Locked-rotor torque (minimum torque at standstill and full voltage) is not less than 100% full-load for 2- and 4-pole motors, 200 hp and less; 40 to 75% for larger 2-pole motors; 50 to 125% for larger 4-pole motors.

CLASS C features high starting torque (locked-rotor over 200%), low starting current. Breakdown torque not less than 190% full-load torque. Slip at full load is between 1½ and 3%.

CLASS D have high slip, high starting torque, low starting current; are used on loads with high intermittent peaks. Driven machine usually has high-inertia flywheel. At no load motor has little slip; when peak load is applied, motor slip increases. Speed reduction lets driven machine absorb energy from flywheel rather than power line.

STARTING Full-voltage, across-the-line starting is used where power supply permits and full-voltage torque and acceleration are not objectionable. Reduction in starting kVA cuts locked-rotor and accelerating torques.

WOUND-ROTOR MOTORS

The wound-rotor (slip-ring) induction motor's rotor winding connects through slip-rings to an external resistance that is cut in and out by a controller.

RESISTANCE vs TORQUE: Resistance of rotor winding affects torque developed at any speed. A high-resistance rotor gives high starting torque with low starting current. But low slip at full load, good efficiency and moderate rotor heating takes a low-resistance rotor. Left-hand curves show rotor-resistance effect on torque. With all resistance in, R_1, full- load starting torque is developed at less than 150% full-load current. Successively shorting out steps, at standstill, develops about 225% full-load torque at R_4. Cutting out more reduces standstill torque. Motor operates like a squirrel-cage motor when all resistance is shorted out.

SPEED CONTROL: Having resistance left in, decreases speed regulation. Righthand curves for a typical motor show that with only two steps shorted out, motor operates at 65% synchronous speed because motor torque equals load torque at that speed. But if load torque drops to 50%, motor shoots forward to about 65% synchronous speed.

However, slip rings are normally shorted after motor comes up to speed. Or, for short-time peak loads, motor is operated with a step or two of resistance cut in. At light loads, motor runs near synchronous speed. When peak loads come on, speed drops; flywheel effect of motor and load cushions power supply from load peak.

OTHER FEATURES: In addition to high-starting-torque, low-starting-current applications, wound-rotor motors are used (1) for high-inertia loads where high slip losses that would have to be dissipated in the rotor of a squirrel-cage motor, in coming up to speed, can be given off as heat in wound-rotor's external resistance (2) where frequent starting, stopping and speed control are needed (3) for continuous operation at reduced speed. (Example: boiler draft fan - combustion control varies external resistance to adjust speed, damper regulates air flow between step speeds and below 50% speed.)

CONTROLS used are across-the-line starters with proper protection (fuses, breakers, etc.) arid secondary control with 5 to 7 resistance steps.

SYNCHRONOUS MOTORS

Synchronous motors run at a fixed or synchronous speed determined by line frequency and number of poles in machine. (Rpm=120xfrequency/number or poles.) Speed is kept constant by locking action of an externally excited DC field. Efficiency is 1 to 3% higher than that of same-size-and-speed induction or DC motors. Also synchronous motors can be operated at power factors from 1.0 down to 0.2 leading for plant power-factor correction. Standard ratings are 1.0 and 0.9 leading PF; machines rated down near 0.2 leading are called synchronous condensers.

STARTING: Pure synchronous motors are not self-starting; so in practice they are built with damper or amortisseur windings. With the field coil shorted through discharge resistor, damper winding acts like a squirrel-cage rotor to bring motor practically to synchronous speed; then field is applied and motor pulls into synchronism, providing motor has developed sufficient "pull-in" torque. Once in synchronism, motor keeps constant speed as long as load torque does not exceed maximum or "pull-out" torque; then machine drops out of synchronism. Driven machine is usually started without load. Low-speed motors may be direct connected.

FIELD AND PF: While motors are rated for specific power factors ... at constant power, increasing DC field current causes power factor to lead, decreasing field tends to make PF lag. But either case increases copper losses.

TYPES: Polyphase synchronous motors in general use are: (1) high-speed motors, 500 rpm up, (a) general-purpose, 500 to 1800 rpm, 200 hp and below, (b) high-speed units over 200, hp including most 2-pole motors (2) low-speed motors below 500 rpm, (3) special high-torque motors.

COSTS: Small high-speed synchronous motors cost more than comparable induction motors, but with low-speed and large high-speed motor, costs favor the synchronous motor. Cost of leading-PF motors increases approximately inversely proportional to the decrease from unity power factor.

DC MOTORS

Chief reason for using DC motors, assuming normal power source is AC, lies in the wide and economical ranges possible of speed control and starting torques. But for constant-speed service, AC motors are generally preferred because they are more rugged and have lower first cost.

STARTING TORQUE: With a shunt motor, torque is proportional to armature current, because field flux remains practically constant for a given setting of the field rheostat. However, the flux of a series field is affected by the current through it. At light loads, flux varies directly with a current, so torque varies as the square of the current. The compound motor (usually cumulative) lies in between the shunt and series motors as to torque.

Upper limit of current input on starting is usually 1.5 to 2 times full-load current to avoid over-heating the commutator, excessive fedder drops or peaking generator. Shunt-motor starting boxes usually allow 125% current at first notch. So motor can develop 125% starting torque. Series motors can develop higher starting torques at same current, since torque increases as current squared. Compound motors develop starting torques higher than shunt motors according to amount of compounding.

SPEED CONTROL: Shunt motor speeds drop only slightly (5% or less) from no load to full load. Decreasing field current raises speed; increasing field reduces speed. But speed is still practically constant for any one field setting. Speed can be controlled by resistance in the armature circuit but regulation is poor.

Series motor speeds decrease much more with increased load, and, conversely, begin to race at low loads, dangerously so if load is completely removed. Speed can be reduced by adding resistance into the armature circuit, increased by shunting the series filed with resistance or short-circuiting series turns.

Compound motors have less constant speed than shunt motors and can be controlled by shunt-field rheostat.

SUMMARY OF MOTOR APPLICATIONS

DC and Single-Phase Motors

Speed Regulation	Speed Control	Starting Torque	Pull-out Torque	Applications
Series				
Varies inversely as the load. Races on light loads and full voltage	Zero to maximum depending on control and load	High. Varies as square of the voltage. Limited by commutation, heating and line capacity	High. Limited by commutation, heating and line capacity	Where high starting torque is required and speed can be regulated. Traction, bridges, hoists, gates, car dumpers, car retarders, etc.
Shunt				
Drops 3 to 5% from no load to full load	Any desired range depending on motor design and type of system.	Good. With constant field, varies directly as voltage applied to armature	High. Limited by commutation, heating and line capacity	Where constant or adjustable speed is required and starting conditions are not severe. Fan, blowers, centrifugal pumps, conveyers, wood working machines, metal working machines, elevators
Compound				
Drops 7 to 20% from no load to full load depending on amount of compounding	Any desired range depending on motor design and type of control	Higher than for shunt, depending on amount of compounding	High. Limited by commutation, heating and line capacity	Where high starting torque combined with fairly constant speed is required. Plunger pumps, punch presses, shears, bending rolls, geared elevators, conveyors, hoists

SUMMARY OF MOTOR APPLICATIONS (cont.)

DC and Single-Phase Motors

Speed Regulation	Speed Control	Starting Torque	Pull-out Torque	Applications
CAPACITOR				
Drops 5% for large to 10% for small sizes	None	150 to 350% of full load depending upon design and size	150% for large to 200% for small sizes	Constant speed service for any starting duty, and quiet operation, where polyphase current cannot be used
COMMUTATOR-TYPE				
Drops 5% for large to 10% for small sizes	Repulsion-induction, none. Brush shifting types 4 to 1 at full load	250% for large to 350% for small sizes	150% for large to 250% for small sizes	Constant speed service for any starting duty, where speed control is required and polyphase current cannot be used
SPLIT-PHASE				
Drops about 10% from no load to full load	None	75% for large to 175% for small sizes	150% for large to 200% for small sizes	Constant speed service where starting is easy. Small fans, centrifugal pumps, and light running machines, where polyphase current is not available

SUMMARY OF MOTOR APPLICATIONS (cont.)

2- and 3-Phase Motors

Speed Regulation	Speed Control	Starting Torque	Pull-out Torque	Applications
GENERAL-PURPOSE SQUIRREL-CAGE (Class B)				
Drops about 3% for large to 5% for small sizes	None, except multi-speed types designed for 2 to 4 fixed speeds	200% of full load for 2-pole to 105% for 16 pole designs	200% of full load	Constant-speed service where starting torque is not excessive. Fans, blowers, rotary compressers, centrifugal pumps
HIGH TORQUE SQUIRREL-CAGE (Class C)				
Drops about 3% for large to 6% for small sizes	None, except multi-speed types designed for 2 to 4 fixed speeds	250% of full load for high speed to 200% for low speed designs	200% of full load	Constant-speed service where fairly high starting torque is required at infrequent intervals with starting current of about 400% of full load. Reciprocating pumps and compressors, crushers, etc.
HIGH SLIP SQUIRREL-CAGE (Class D)				
Drops about 10 to 15% from no load to full load	None, except multi-speed types designed for 2 to 4 fixed speeds	225 to 300% of full load, depending on speed with rotor resistance	200% Will usually not stall until loaded to Max. torque, which occurs at standstill	Constant-speed service and high starting torque, if starting is not too frequent, and for taking high peak loads with or without flywheels. Punch presses, shears and elevators, etc.

SUMMARY OF MOTOR APPLICATIONS (cont.)

2- and 3-Phase Motors

Speed Regulation	Speed Control	Starting Torque	Pull-out Torque	Applications
LOW TORQUE SQUIRREL-CAGE (Class F)				
Drops about 3% for large to 5% for small sizes	None, except multi-speed types designed for 2 to 4 fixed speeds	50% of full load for high speed to 90% for low speed designs	150 to 170% of full load	Constant speed service where starting duty is light. Fans blowers, centrifugal pumps and similar loads
WOUND-ROTOR				
With rotor rings short circuited, drops about 3% for large to 5% for small sizes	Speed can be reduced to 50% by rotor resistance to obtain stable operation. Speed varies inversely as load	Up to 300% depending on external resistance in rotor circuit and how distributed	200%. when rotor slip rings are short circuited	Where high starting torque with low starting current or where limited speed control is required. Fans, centrifugal and plunger pumps, compressors, conveyers, hoists, cranes
SYNCHRONOUS				
Constant	None, except special motors designed for 2 fixed speeds	40% for slow to 160% for medium speed 80% PF designs. Special designs develop higher torques	Unity-of motors 170%; 80% PF motors 225%. Special designs up to 300%	For constant-speed service, direct connection to slow speed machines and where power factor correction is required.

MOTOR FRAME DIMENSIONS

Frame #	Dimension in Inches — NEMA						
	D	E	F	U	V	M+N	Keyway
445U	11	9	8-1/4	2-7/8	8-3/8	24-3/8	3/4x3/8
445T	11	9	8-1/4	3-3/8	8-1/4	24-1/4	7/8x7/16
444U	11	9	7-1/4	2-7/8	8-3/8	23-3/8	3/4x3/8
444T	11	9	7-1/4	3-3/8	8-1/4	23-1/4	7/8x7/16
405U	10	8	6-7/8	2-3/8	6-7/8	20-5/8	5/8x5/16
405T	10	8	6-7/8	2-7/8	7	20-3/4	3/4x3/8
404U	10	8	6-1/8	2-3/8	6-7/8	19-7/8	5/8x5/16
404T	10	8	6-1/8	2-7/8	7	20	3/4x3/8
365U	9	7	6-1/8	2-1/8	6-1/8	18-3/8	1/2x1/4
365T	9	7	6-1/8	2-3/8	5-5/8	17-7/8	5/8x5/16
364U	9	7	5-5/8	2-1/8	6-1/8	17-7/8	1/2x1/4
364T	9	7	5-5/8	2-3/8	5-5/8	17-3/8	5/8x5/16
326U	8	6-1/4	6	1-7/8	5-3/8	16-7/8	1/2x1/4
326TS	8	6-1/4	6	1-7/8	3-1/2	15	1/2x1/4
326T	8	6-1/4	6	2-1/8	5	16-1/2	1/2x1/4
324U	8	6-1/4	5-1/4	1-7/8	5-3/8	16-1/8	1/2x1/4
324T	8	6-1/4	5-1/4	2-1/8	5	15-3/4	1/2x1/4
286U	7	5-1/2	5-1/2	1-5/8	4-5/8	15-1/8	3/8x3/16
286T	7	5-1/2	5-1/2	1-7/8	4-3/8	14-7/8	1/2x1/4
284U	7	5-1/2	4-3/4	1-5/8	4-5/8	14-3/8	3/8x3/16
284TS	7	5-1/2	4-3/4	1-5/8	3	12-3/4	3/8x3/16
284T	7	5-1/2	4-3/4	1-7/8	4-3/8	14-1/8	1/2x1/4
256U	6-1/4	5	5	1-3/8	3-1/2	13	5/16x5/32
256T	6-1/4	5	5	1-5/8	3-3/4	13-1/4	3/8x3/16
254U	6-1/4	5	4-1/8	1-3/8	3-1/2	12-1/8	5/16x5/32
254T	6-1/4	5	4-1/8	1-5/8	3-3/4	12-3/8	3/8x3/16
215T	5-1/4	4-1/4	3-1/2	1-3/8	3-1/8	10-3/8	5/16x5/32
215	5-1/4	4-1/4	3-1/2	1-1/8	2-3/4	10	1/4x1/8
213T	5-1/4	4-1/4	2-3/4	1-3/8	3-1/8	9-5/8	5/16x5/32
213	5-1/4	4-1/4	2-3/4	1-1/8	2-3/4	9-1/4	1/4x1/8
184T	4-1/2	3-3/4	2-3/4	1-1/8	2-1/2	8-1/4	1/4x1/8
184	4-1/2	3-3/4	2-3/4	7/8	2	7-3/4	3/16x3/32
182T	4-1/2	3-3/4	2-1/4	1-1/8	2-1/2	7-3/4	1/4x1/8
182	4-1/2	3-3/4	2-1/4	7/8	2	7-1/4	3/16x3/32
145T	3-1/2	2-3/4	2-1/2	7/8	2	7	3/16x3/32
143T	3-1/2	2-3/4	2	7/8	2	6-1/2	3/16x3/32
66	4-1/8	2-15/16	2-1/2	3/4		7-7/8	3/16x3/32
56	3-1/2	2-7/16	2-1/2	5/8		6-1/8	3/16x3/32
48	2	2-1/8	1-3/8	1/2		5-3/8	
42	2-5/8	1-3/4	27/32	3/8		4-1/32	

Standards established by National Electrical Manufactures Association.

FRONTAL VIEW OF TYPICAL MOTOR

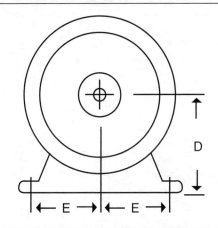

REFERENCE PAGE 6-14 FOR DIMENSIONS

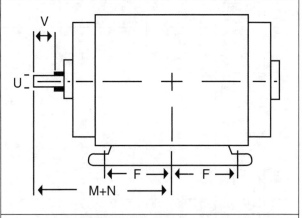

SIDE VIEW OF TYPICAL MOTOR

MOTOR FRAME DIMENSIONS

Frame No.	Shaft U	Shaft V	Key W	Key T	Key L	Dimensions—Inches A	B	D	E	F	BA
48	1/2	1 1/2*	flat	3/64	—	5 5/8*	3 1/2*	3	2 1/8	1 3/8	2 1/2
56	5/8	1 7/8*	3/16	3/16	1 3/8	6 1/2*	4 1/4*	3 1/2	2 7/16	1 1/2	2 3/4
143T	7/8	2	3/16	3/16	1 3/8	7	6	3 1/2	2 3/4	2	2 1/4
145T	7/8	2	3/16	3/16	1 3/8	7	6	3 1/2	2 3/4	2 1/2	2 1/4
182	7/8	2	3/16	3/16	1 3/8	9	6 1/2	4 1/2	3 3/4	2 1/4	2 3/4
182T	1 1/8	2 1/2	1/4	1/4	1 3/4	9	6 1/2	4 1/2	3 3/4	2 1/4	2 3/4
184	7/8	2	3/16	3/16	1 3/8	9	7 1/2	4 1/2	3 3/4	2 3/4	2 3/4
184T	1 1/8	2 1/2	1/4	1/4	1 3/4	9	7 1/2	4 1/2	3 3/4	2 3/4	2 3/4
203	3/4	2	3/16	3/16	1 3/8	10	7 1/2	5	4	2 3/4	3 1/8
204	3/4	2	3/16	3/16	1 3/8	10	8 1/2	5	4	3 1/4	3 1/8
213	1 1/8	2 3/4	1/4	1/4	2	10 1/2	7 1/2	5 1/4	4 1/4	2 3/4	3 1/2
213T	1 3/8	3 1/4	5/16	5/16	2 3/8	10 1/2	7 1/2	5 1/4	4 1/4	2 3/4	3 1/2
215	1 1/8	2 3/4	1/4	1/4	2	10 1/2	9	5 1/4	4 1/4	3 1/2	3 1/2
215T	1 3/8	3 1/8	5/16	5/16	2 3/8	10 1/2	9	5 1/4	4 1/4	3 1/2	3 1/2
224	1	2 3/4	1/4	1/4	2	11	8 3/4	5 1/2	4 1/2	3 3/8	3 1/2
225	1	2 3/4	1/4	1/4	2	11	9 1/2	5 1/2	4 1/2	3 3/4	3 1/2
254	1 1/8	3 1/8	1/4	1/4	2 3/8	12 1/2	10 3/4	6 1/4	5	4 1/8	4 1/4
254U	1 3/8	3 1/2	5/16	5/16	2 3/4	12 1/2	10 3/4	6 1/4	5	4 1/8	4 1/4
254T	1 5/8	3 3/4	3/8	3/8	2 7/8	12 1/2	10 3/4	6 1/4	5	4 1/8	4 1/4
256U	1 3/8	3 1/2	5/16	5/16	2 3/4	12 1/2	12 1/2	6 1/4	5	5	4 1/4
256T	1 5/8	3 3/4	3/8	3/8	2 7/8	12 1/2	12 1/2	6 1/4	5	5	4 1/4
284	1 1/4	3 1/2	1/4	1/4	2 3/4	14	12 1/2	7	5 1/2	4 3/4	4 3/4
284U	1 5/8	4 5/8	3/8	3/8	3 3/4	14	12 1/2	7	5 1/2	4 3/4	4 3/4
284T	1 7/8	4 3/8	1/2	1/2	3 1/4	14	12 1/2	7	5 1/2	4 3/4	4 3/4
284TS	1 5/8	3	3/8	3/8	1 7/8	14	12 1/2	7	5 1/2	4 3/4	4 3/4
286U	1 5/8	4 5/8	3/8	3/8	3 3/4	14	14	7	5 1/2	5 1/2	4 3/4
286T	1 7/8	4 3/8	1/2	1/2	3 1/4	14	14	7	5 1/2	5 1/2	4 3/4
286TS	1 5/8	3	3/8	3/8	1 7/8	14	14	7	5 1/2	5 1/2	4 3/4
324	1 5/8	4 5/8	3/8	3/8	3 3/4	16	14	8	6 1/4	5 1/4	5 1/4
324U	1 7/8	5 3/8	1/2	1/2	4 1/4	16	14	8	6 1/4	5 1/4	5 1/4
324S	1 5/8	3	3/8	3/8	1 7/8	16	14	8	6 1/4	5 1/4	5 1/4
324T	2 1/8	5	1/2	1/2	3 7/8	16	14	8	6 1/4	5 1/4	5 1/4
324TS	1 7/8	3 1/2	1/2	1/2	2	16	14	8	6 1/4	5 1/4	5 1/4
326	1 5/8	4 5/8	3/8	3/8	3 3/4	16	15 1/2	8	6 1/4	6	5 1/4
326U	1 7/8	5 3/8	1/2	1/2	4 1/4	16	15 1/2	8	6 1/4	6	5 1/4
326S	1 5/8	3	3/8	3/8	1 7/8	16	15 1/2	8	6 1/4	6	5 1/4
326T	2 1/8	5	1/2	1/2	3 7/8	16	15 1/2	8	6 1/4	6	5 1/4
326TS	1 7/8	3 1/2	1/2	1/2	2	16	15 1/2	8	6 1/4	6	5 1/4

*Not NEMA standard dimensions

MOTOR FRAME DIMENSIONS (cont.)

Frame No.	Shaft U	Shaft V	Key W	Key T	Key L	A	B	D	E	F	BA
364	1 7/8	5 3/8	1/2	1/2	4 1/4	18	15 1/4	9	7	5 5/8	5 7/8
364S	1 5/8	3	3/8	3/8	1 7/8	18	15 1/4	9	7	5 5/8	5 7/8
364U	2 1/8	6 1/8	1/2	1/2	5	18	15 1/4	9	7	5 5/8	5 7/8
364US	1 7/8	3 1/2	1/2	1/2	2	18	15 1/4	9	7	5 5/8	5 7/8
364T	2 3/8	5 5/8	5/8	5/8	4 1/4	18	15 1/4	9	7	5 5/8	5 7/8
364TS	1 7/8	3 1/2	1/2	1/2	2	18	15 1/4	9	7	5 5/8	5 7/8
365	1 7/8	5 3/8	1/2	1/2	4 1/4	18	16 1/4	9	7	6 1/8	5 7/8
365S	1 5/8	3	3/8	3/8	1 7/8	18	16 1/4	9	7	6 1/8	5 7/8
365U	2 1/8	6 1/8	1/2	1/2	5	18	16 1/4	9	7	6 1/8	5 7/8
365US	1 7/8	3 1/2	1/2	1/2	2	18	16 1/4	9	7	6 1/8	5 7/8
365T	2 3/8	5 5/8	5/8	5/8	4 1/4	18	16 1/4	9	7	6 1/8	5 7/8
365TS	1 7/8	3 1/2	1/2	1/2	2	18	16 1/4	9	7	6 1/8	5 7/8
404	2 1/8	6 1/8	1/2	1/2	5	20	16 1/4	10	8	6 1/8	6 5/8
404S	1 7/8	3 1/2	1/2	1/2	2	20	16 1/4	10	8	6 1/8	6 5/8
404U	2 3/8	6 7/8	5/8	5/8	5 1/2	20	16 1/4	10	8	6 1/8	6 5/8
404US	2 1/8	4	1/2	1/2	2 3/4	20	16 1/4	10	8	6 1/8	6 5/8
404T	2 7/8	7	3/4	3/4	5 5/8	20	16 1/4	10	8	6 1/8	6 5/8
404TS	2 1/8	4	1/2	1/2	2 3/4	20	16 1/4	10	8	6 1/8	6 5/8
405	2 1/8	6 1/8	1/2	1/2	5	20	17 3/4	10	8	6 7/8	6 5/8
405S	1 7/8	3 1/2	1/2	1/2	2	20	17 3/4	10	8	6 7/8	6 5/8
405U	2 3/8	6 7/8	5/8	5/8	5 1/2	20	17 3/4	10	8	6 7/8	6 5/8
405US	2 1/8	4	1/2	1/2	2 3/4	20	17 3/4	10	8	6 7/8	6 5/8
405T	2 7/8	7	3/4	3/4	5 5/8	20	17 3/4	10	8	6 7/8	6 5/8
405TS	2 1/8	4	1/2	1/2	2 3/4	20	17 3/4	10	8	6 7/8	6 5/8
444	2 3/8	6 7/8	5/8	5/8	5 1/2	22	18 1/2	11	9	7 1/4	7 1/2
444S	2 1/8	4	1/2	1/2	2 3/4	22	18 1/2	11	9	7 1/4	7 1/2
444U	2 7/8	8 3/8	3/4	3/4	7	22	18 1/2	11	9	7 1/4	7 1/2
444US	2 1/8	4	1/2	1/2	2 3/4	22	18 1/2	11	9	7 1/4	7 1/2
444T	3 3/8	8 1/4	7/8	7/8	6 7/8	22	18 1/2	11	9	7 1/4	7 1/2
444TS	2 3/8	4 1/2	5/8	5/8	3	22	18 1/2	11	9	7 1/4	7 1/2
445	2 3/8	6 7/8	5/8	5/8	5 1/2	22	20 1/2	11	9	8 1/4	7 1/2
445S	2 1/8	4	1/2	1/2	2 3/4	22	20 1/2	11	9	8 1/4	7 1/2
445U	2 7/8	8 3/8	3/4	3/4	7	22	20 1/2	11	9	8 1/4	7 1/2
445US	2 1/8	4	1/2	1/2	2 3/4	22	20 1/2	11	9	8 1/4	7 1/2
445T	3 3/8	8 1/4	7/8	7/8	6 7/8	22	20 1/2	11	9	8 1/4	7 1/2
445TS	2 3/8	4 1/2	5/8	5/8	3	22	20 1/2	11	9	8 1/4	7 1/2
504U	2 7/8	8 3/8	3/4	3/4	7 1/4	25	21	12 1/2	10	8	8 1/2
504S	2 1/8	4	1/2	1/2	2 3/4	25	21	12 1/2	10	8	8 1/2
505	2 7/8	8 3/8	3/4	3/4	7 1/4	25	23	12 1/2	10	9	8 1/2
505S	2 1/8	4	1/2	1/2	2 3/4	25	23	12 1/2	10	9	8 1/2

MOTOR FRAME TABLE

Frame No. Series	Third/Fourth Digit of Frame No.						
	D	1	2	3	4	5	6	7
140	3.50	3.00	3.50	4.00	4.50	5.00	5.50	6.25
160	4.00	3.50	4.00	4.50	5.00	5.50	6.25	7.00
180	4.50	4.00	4.50	5.00	5.50	6.25	7.00	8.00
200	5.00	4.50	5.00	5.50	6.50	7.00	8.00	9.00
210	5.25	4.50	5.00	5.50	6.25	7.00	8.00	9.00
220	5.50	5.00	5.50	6.25	6.75	7.50	9.00	10.00
250	6.25	5.50	6.25	7.00	8.25	9.00	10.00	11.00
280	7.00	6.25	7.00	8.00	9.50	10.00	11.00	12.50
320	8.00	7.00	8.00	9.00	10.50	11.00	12.00	14.00
360	9.00	8.00	9.00	10.00	11.25	12.25	14.00	16.00
400	10.00	9.00	10.00	11.00	12.25	13.75	16.00	18.00
440	11.00	10.00	11.00	12.50	14.50	16.50	18.00	20.00
500	12.50	11.00	12.50	14.00	16.00	18.00	20.00	22.00
580	14.50	12.50	14.00	16.00	18.00	20.00	22.00	25.00
680	17.00	16.00	18.00	20.00	22.00	25.00	28.00	32.00

MOTOR FRAME TABLE (cont.)

Frame No. Series	D	Third/Fourth Digit of Frame No.							
		8	9	10	11	12	13	14	15
140	3.50	7.00	8.00	9.00	10.00	11.00	12.50	14.00	16.00
160	4.00	8.00	9.00	10.00	11.00	12.50	14.00	16.00	18.00
180	4.50	9.00	10.00	11.00	12.50	14.00	16.00	18.00	20.00
200	5.00	10.00	11.00	—	—	—	—	—	—
210	5.25	10.00	11.00	12.50	14.00	16.00	18.00	20.00	22.00
220	5.50	11.00	12.50	—	—	—	—	—	—
250	6.25	12.50	14.00	16.00	18.00	20.00	22.00	25.00	28.00
280	7.00	14.00	16.00	18.00	20.00	22.00	25.00	28.00	32.00
320	8.00	16.00	18.00	20.00	22.00	25.00	28.00	32.00	36.00
360	9.00	18.00	20.00	22.00	25.00	28.00	32.00	36.00	40.00
400	10.00	20.00	22.00	25.00	28.00	32.00	36.00	40.00	45.00
440	11.00	22.00	25.00	28.00	32.00	36.00	40.00	45.00	50.00
500	12.50	25.00	28.00	32.00	36.00	40.00	45.00	50.00	56.00
580	14.50	28.00	32.00	36.00	40.00	45.00	50.00	56.00	63.00
680	17.00	36.00	40.00	45.00	50.00	56.00	63.00	71.00	80.00

MOTOR FRAME LETTERS

LETTER	DESIGNATION
G	Gasoline pump motor
K	Sump pump motor
M and N	Oil burner motor
S	Standard short shaft for direct connection
T	Standard dimensions established
U	Previously used as frame designation for which standard dimensions are established
Y	Special mounting dimensions required from manufacturer
Z	Standard mounting dimensions except shaft extension

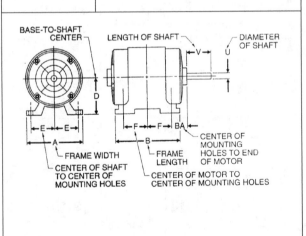

BASE-TO-SHAFT CENTER

LENGTH OF SHAFT

DIAMETER OF SHAFT

V

U

D

E — E

F — F — BA

A

B

CENTER OF MOUNTING HOLES TO END OF MOTOR

FRAME WIDTH

CENTER OF SHAFT TO CENTER OF MOUNTING HOLES

FRAME LENGTH

CENTER OF MOTOR TO CENTER OF MOUNTING HOLES

TYPES OF ENCLOSURES

OPEN-TYPE
has full openings in frame and endbells for maximum ventilation, is lowest cost enclosure.

SEMI-PROTECTED
has screens in top openings to keep out falling objects. PROTECTED has screens in bottom too.

DRIP-PROOF
has upper parts covered to keep out drippings falling at angle not over 15° from vertical.

SPLASH-PROOF
is baffled at bottom to keep out particles coming at angle not over 100° from vertical.

TOTALLY-ENCLOSED
can be non-ventilated, separately ventilated, or explosion proof for hazardous atmospheres.

FAN-COOLED
totally-enclosed motor has double covers. Fan, behind vented outer shroud, is run by motor.

INSULATION, TEMPERATURE

CLASS A (cotton, silk, paper or other organics impregnated with insulating varnish) is considered standard for most applications, allows 105° C total temperature.

40°C ambient
40°C rise by thermometer
15°C "hot-spot" allowance
10°C service factor
—————————————
105°C total temperature

CLASS B (mica, asbestos, fiber-glass, other inorganics) allows 130°C total temperature.

40°C ambient
70°C rise by thermometer
20°C "hot-spot" allowance
—————————————
130°C total temperature

CLASS H (including silicone family) is for special high-temperature applications.

SPECIAL CLASS A is highly resistant, but not "proof," against severe moisture, dampness; conductive, corrosive or abrasive dusts and vapors.

TROPICAL is for excessive moisture, high ambients, corrosion, fungus, vermin, insects.

SHAFT COUPLING SELECTIONS

Coupling number	Rated torque (lb.-in.)	Maximum shock torque (lb.-in.)
10-101-A	16	45
10-102-A	36	100
10-103-A	80	220
10-104-A	132	360
10-105-A	176	480
10-106-A	240	660
10-107-A	325	900
10-108-A	525	1450
10-109-A	875	2450
10-110-A	1250	3500
10-111-A	1800	5040
10-112-A	2200	6160

V-BELTS

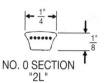

NO. 0 SECTION
"2L"

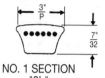

NO. 1 SECTION
"3L"

NO. 2 SECTION
"4L"
A

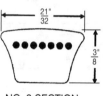

NO. 3 SECTION
"5 L"
B

V-BELT/MOTOR SIZE

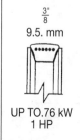

$\frac{3"}{8}$
9.5. mm

UP TO .76 kW
1 HP

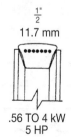

$\frac{1"}{2}$
11.7 mm

.56 TO 4 kW
5 HP

$\frac{21"}{32}$
16.7 mm

2.5 kW AND UP
3 HP

TYPICAL MOTOR EFFICIENCIES

HP	Standard Motor (%)	Energy-Efficient Motor (%)
1	76.5	84.0
1.5	78.5	85.5
2	79.9	86.5
3	80.8	88.5
5	83.1	88.6
7.5	83.8	90.2
10	85.0	90.3
15	86.5	91.7
20	87.5	92.4
25	88.0	93.0
30	88.1	93.1
40	89.3	93.6
50	90.4	93.7
75	90.8	95.0
100	91.6	95.4
125	91.8	95.8
150	92.3	96.0
200	93.3	96.1
250	93.6	96.2
300	93.8	96.5

COMMON SERVICE FACTORS

Equipment	Service factor
Blowers	
Centrifugal	1.00
Vane	1.25
Compressors	
Centrifugal	1.25
Vane	1.50
Conveyors	
Uniformly loaded or fed	1.50
Heavy-duty	2.00
Elevators	
Bucket	2.00
Freight	2.25
Extruders	
Plastic	2.00
Metal	2.50
Fans	
Light-duty	1.00
Centrifugal	1.50
Machine tools	
Bending roll	2.00
Punch press	2.25
Tapping machine	3.00
Mixers	
Concrete	2.00
Drum	2.25
Paper mills	
De-barking machines	3.00
Beater and pulper	2.00
Bleacher	1.00
Dryers	2.00
Log haul	2.00
Printing presses	1.50
Pumps	
Centrifugal—general	1.00
Centrifugal—sewage	2.00
Reciprocating	2.00
Rotary	1.50
Textile	
Batchers	1.50
Dryers	1.50
Looms	1.75
Spinners	1.50
Woodworking machines	1.00

DC MOTOR PERFORMANCE CHARACTERISTICS

Performance Characteristics	Voltage 10% below Rated Voltage		Voltage 10% above Rated Voltage	
	Shunt	Compound	Shunt	Compound
Starting Torque	−15%	−15%	+15%	+15%
Speed	−5%	−6%	+5%	+6%
Current	+12%	+12%	−8%	−8%
Field Temperature	Increases	Decreases	Increases	Increases
Armature Temperature	Increases	Increases	Decreases	Decreases
Commutator Temperature	Increases	Increases	Decreases	Decreases

MAXIMUM ACCELERATION TIME

Frame Number	Maximum Acceleration Time (in seconds)
48 and 56	8
143-286	10
324-326	12
364-505	15

SMALL MOTOR GUIDE

For AC type, 115 volt, 60 Hz, single-phase

HP	Amps @ Full Load	RPM Speed
2	19.0 to 23.0	3450
1-1/2	19.6	1725
	16.4 to 19.6	3450
1	13.6 to 16.0	1725
	13.0 to 15.0	3450
3/4	9.5	1075
	11.6	1725
	11.8	3450
1/2	7.3	1075
	7.0 to 9.2	1725
	9.8	3450
1/3	5.1	1075
	5.0 to 7.2	1140
	5.3 to 6.8	1725
	5.6 to 6.5	3450
1/4	6.9	850
	3.4 to 6.8	1075
	5.6 to 6.8	1140
	3.1 to 3.6	1625
	4.4 to 6.3	1725
1/6	2.4 to 5.0	1075
	4.0 to 4.9	1140
	4.0 to 4.8	1550
	3.3 to 4.7	1725
1/8	1.8 to 5.0	1075
	3.8	1140
	2.5	1725
1/10	4.0	1050
	3.5	1550
1/12	3.2	850
	4.1	1550
	2.8	1725
1/15	2.8	1550
1/20	2.5	1550

This chart is for small motors used in HVAC equip., fans, small pumps, etc. Specifications may vary per job requirement and motor used. When referencing 230 volt motors, divide the amperes by 2.

THREE-PHASE MOTOR REQUIREMENTS

Motor HP	Amperes at Full Load		Size of Conduit in Inches		Minimum Wire Size Rubber – AWG	
	230 Volts	460 Volts	230 Volts	460 Volts	230 Volts	460 Volts
200	480	240	—	3	—	500 kcmil
150	360	180	4	2-1/2	1000 kcmil	300 kcmil
125	310	155	3-1/2	2-1/2	750 kcmil	0000
100	245	123	3	2	500 kcmil	000
75	180	90	2-1/2	2	0000	0
60	149	75	2-1/2	1-1/2	200 kcmil	1
50	125	63	2	1-1/4	000	3
40	101	51	2	1-1/4	00	4
30	77	39	1-1/2	1-1/4	1	6
25	64	32	1-1/4	1-1/4	3	6
20	52	26	1-1/4	3/4	4	8
15	38	19	1-1/4	3/4	6	10
10	27	14	3/4	1/2	8	12
7-1/2	22	11	3/4	1/2	8	14
5	15	7.5	1/2	1/2	12	14
3	9	4.5	1/2	1/2	14	14
2	6	3	1/2	1/2	14	14
1-1/2	4.7	2.4	1/2	1/2	14	14
1	3.3	1.7	1/2	1/2	14	14

If voltage drop exceeds limits, the wire size should be adjusted.   kcmil = 1000 circular mils

DIRECT CURRENT MOTOR REQUIREMENTS

Motor HP	Amperes at Full Load		Size of Conduit in Inches		Minimum Wire Size Rubber – AWG	
	115 Volts	230 Volts	115 Volts	230 Volts	115 Volts	230 Volts
100	—	355	—	4	—	1000 kcmil
75	—	268	—	3-1/2	—	600 kcmil
60	—	215	—	3	—	400 kcmil
50	360	180	4	2-1/2	1000 kcmil	300 kcmil
40	292	146	3-1/2	2-1/2	700 kcmil	0000
30	220	110	3	2	400 kcmil	00
25	184	92	2-1/2	2	300 kcmil	0
20	140	74	2	1-1/2	000	1
15	112	56	2	1-1/4	00	4
10	75	38	1-1/2	1	1	6
7-1/2	58	28.7	1-1/4	1	3	6
5	40	19.8	1	3/4	6	10
3	23	12.3	3/4	1/2	8	12
2	16.1	8.3	3/4	1/2	10	14
1-1/2	12.5	6.3	1/2	1/2	12	14
1	8.4	4.2	1/2	1/2	14	14

Kcmil = 1000 circular mils

MOTOR TORQUE (INCH POUNDS-FORCE)

$$\frac{63025 \times HP}{RPM} = \text{Torque in inch pounds - force}$$

$$\left(\begin{array}{c} \text{Divide the above by 12} \\ \text{to obtain torque in foot pounds} \end{array} \right)$$

Motor	RPM Speeds of Motors			
HP	68	100	155	190
600	556,103	378,150	243,968	199,026
550	509,761	346,638	223,637	182,441
500	463,419	315,125	203,306	165,855
450	417,077	283,613	182,976	149,270
400	370,735	252,100	162,645	132,684
350	324,393	220,588	142,315	116,099
300	278,051	189,075	121,984	99,513
275	254,881	173,319	111,819	91,220
250	231,710	157,563	101,653	82,928
225	208,539	141,806	91,488	74,635
200	185,368	126,050	81,323	66,342
175	162,197	110,294	71,157	58,049
150	139,026	94,538	60,992	49,757
125	115,855	78,781	50,827	41,464
100	92,684	63,025	40,661	33,171
90	83,415	56,723	36,595	29,854
80	74,147	50,420	32,529	26,537
70	64,879	44,118	28,463	23,220
60	55,610	37,815	24,397	19,903
50	46,342	31,513	20,331	16,586
40	37,074	25,210	16,265	13,268
30	27,805	18,908	12,198	9,951
25	23,171	15,756	10,165	8,293
20	18,537	12,605	8,132	6,634
15	13,903	9,454	6,099	4,976
10	9,268	6,303	4,066	3,317
7-1/2	6,951	4,727	3,050	2,488
5	4,634	3,151	2,033	1,659
3	2,781	1,891	1,220	995
2	1,854	1,261	813	663
1-1/2	1,390	945	610	498
1	927	630	407	332

MOTOR TORQUE (INCH POUNDS-FORCE) (cont.)

$$\frac{63025 \times HP}{RPM} = \text{Torque in inch pounds - force}$$

$$\left(\begin{array}{c} \text{Divide the above by 12} \\ \text{to obtain torque in foot pounds} \end{array} \right)$$

Motor	RPM Speeds of Motors			
HP	500	750	850	1000
600	75,630	50,420	44,488	37,815
550	69,328	46,218	40,781	34,664
500	63,025	42,017	37,074	31,513
450	56,723	37,815	33,366	28,361
400	50,420	33,613	29,659	25,210
350	44,118	29,412	25,951	22,059
300	37,815	25,210	22,244	18,908
275	34,664	23,109	20,390	17,332
250	31,513	21,008	18,537	15,756
225	28,361	18,908	16,683	14,181
200	25,210	16,807	14,829	12,605
175	22,059	14,706	12,976	11,029
150	18,908	12,605	11,122	9,454
125	15,756	10,504	9,268	7,878
100	12,605	8,403	7,415	6,303
90	11,345	7,563	6,673	5,672
80	10,084	6,723	5,932	5,042
70	8,824	5,882	5,190	4,412
60	7,563	5,042	4,449	3,782
50	6,303	4,202	3,707	3,151
40	5,042	3,361	2,966	2,521
30	3,782	2,521	2,224	1,891
25	3,151	2,101	1,854	1,576
20	2,521	1,691	1,483	1,261
15	1,891	1,261	1,112	945
10	1,261	840	741	630
7-1/2	945	630	556	473
5	630	420	371	315
3	378	252	222	189
2	252	168	148	126
1-1/2	189	126	111	95
1	126	84	74	63

MOTOR TORQUE (INCH POUNDS-FORCE) (cont.)

$$\frac{63025 \times HP}{RPM} = \text{Torque in inch pounds - force}$$

$$\left(\begin{array}{c} \text{Divide the above by 12} \\ \text{to obtain torque in foot pounds} \end{array}\right)$$

Motor	RPM Speeds of Motors			
HP	1050	1550	1725	3450
600	36,014	24,397	21,922	10,961
550	33,013	22,364	20,095	10,047
500	30,012	20,331	18,268	9,134
450	27,011	18,298	16,441	8,221
400	24,010	16,265	14,614	7,307
350	21,008	14,231	12,788	6,394
300	18,007	12,198	10,961	5,480
275	16,507	11,182	10,047	5,024
250	15,006	10,165	9,134	4,567
225	13,505	9,149	8,221	4,110
200	12,005	8,132	7,307	3,654
175	10,504	7,116	6,394	3,197
150	9,004	6,099	5,480	2,740
125	7,503	5,083	4,567	2,284
100	6,002	4,066	3,654	1,827
90	5,402	3,660	3,288	1,644
80	4,802	3,253	2,923	1,461
70	4,202	2,846	2,558	1,279
60	3,601	2,440	2,192	1,096
50	3,001	2,033	1,827	913
40	2,401	1,626	1,461	731
30	1,801	1,220	1,096	548
25	1,501	1,017	913	457
20	1,200	813	731	365
15	900	610	548	274
10	600	407	365	183
7-1/2	450	305	274	137
5	300	203	183	91
3	180	122	110	55
2	120	81	73	37
1-1/2	90	61	55	27
1	60	41	37	18

HORSEPOWER TO TORQUE CONVERSION

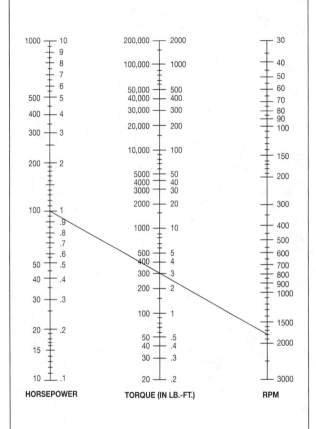

HORSEPOWER

TORQUE (IN LB.-FT.)

RPM

1φ 115 V MOTORS AND CIRCUITS – 120 V SYSTEM

Size of motor HP	Amp	Motor overload protection Low-peak or Fusetron® Motor less than 40°C or greater than 1.15 SF (Max. fuse 125%)	All other motors (Max. fuse 115%)	Switch 115% minimum or HP rated or fuse holder size	Minimum size of starter	Controller termination temperature rating 60°C THW TW		75°C TW THW		Minimum size of copper wire and trade conduit Wire size (AWG or kcmil)	Conduit (inches)
1/6	4.4	5	5	30	00	•	•	•	•	14	1/2
1/4	5.8	7	6 1/4	30	00	•	•	•	•	14	1/2
1/3	7.2	9	8	30	00	•	•	•	•	14	1/2
1/2	9.8	12	10	30	00	•	•	•	•	14	1/2
3/4	13.8	15	15	30	00	•	•	•	•	12	1/2
1	16	20	17 1/2	30	00	•	•	•	•	12	1/2
1 1/2	20	25	20	30	01	•	•	•	•	10	1/2
2	24	30	25	30	01	•	•	•	•	10	1/2

1φ 230 V MOTORS AND CIRCUITS – 240 V SYSTEM

Size of motor HP	Amp	Motor overload protection Low-peak or Fusetron® Motor less than 40°C or greater than 1.15 SF (Max. fuse 125%)	All other motors (Max. fuse 115%)	Switch 115% minimum or HP rated or fuse holder size	Minimum size of starter	Controller termination temperature rating 60°C TW THW	75°C TW THW	Wire size (AWG or kcmil)	Conduit (inches)
1/6	2.2	2 1/2	2 1/2	30	00	• •	• •	14	1/2
1/4	2.9	3 1/2	3 2/10	30	00	• •	• •	14	1/2
1/3	3.6	4 1/2	4	30	00	• •	• •	14	1/2
1/2	4.9	5 6/10	5 6/10	30	00	• •	• •	14	1/2
3/4	6.9	8	7 1/2	30	00	• •	• •	14	1/2
1	8	10	9	30	00	• •	• •	14	1/2
1 1/2	10	12	10	30	00	• •	• •	14	1/2
2	12	15	12	30	0	• •	• •	14	1/2
3	17	20	17 1/2	30	1	• •	• •	12	1/2

6-35

1φ 230 V MOTORS AND CIRCUITS – 240 V SYSTEM (cont.)

Size of motor HP	Amp	Motor overload protection Low-peak or Fusetron®		Switch 115% minimum or HP rated or fuse holder size	Minimum size of starter	Controller termination temperature rating				Minimum size of copper wire and trade conduit	
		Motor less than 40°C or greater than 1.15 SF (Max. fuse 125%)	All other motors (Max. fuse 115%)			60°C		75°C		Wire size (AWG or kcmil)	Conduit (inches)
						TW	THW	TW	THW		
5	28	35	30*	60	2		•			8	3/4
						•				8	1/2
									•	10	1/2
7 1/2	40	50	45	60	2	•	•	•		6	3/4
									•	8	3/4
10	50	60	50	60	3	•	•	•		4	1
									•	6	3/4

* Fuse reducers required.

3φ 230 V MOTORS AND CIRCUITS – 240 V SYSTEM

Size of motor HP	Amp	Motor less than 40°C or greater than 1.15 SF (Max. fuse 125%)	All other motors (Max. fuse 115%)	Switch 115% minimum or HP rated or fuse holder size	Minimum size of starter	60°C TW	60°C THW	75°C TW	75°C THW	Wire size (AWG or kcmil)	Conduit (inches)
1/2	2	2 1/4	2 1/4	30	00		•	•	•	14	1/2
3/4	2.8	3 1/2	3 2/10	30	00	•	•	•	•	14	1/2
1	3.6	4 1/2	4	30	00	•	•	•	•	14	1/2
1 1/2	5.2	6 1/4	5 6/10	30	00	•	•	•	•	14	1/2
2	6.8	8	7 1/2	30	0		•	•	•	14	1/2
3	9.6	12	10	30	0	•	•	•	•	14	1/2
5	15.2	17 1/2	17 1/2	30	1	•	•	•	•	14	1/2
7 1/2	22	25	25	30	1		•	•	•	10	1/2
10	28	35	30*	60	2	•			•	8	3/4
									•	10	1/2
15	42	50	45	60	2	•			•	6	1
									•	6	3/4

* Fuse reducers required.

6-37

3φ 230 V MOTORS AND CIRCUITS – 240 V SYSTEM (cont.)

Size of motor HP	Amp	Motor overload protection Low-peak or Fusetron® — Motor less than 40°C or greater than 1.15 SF (Max. fuse 125%)	All other motors (Max. fuse 115%)	Switch 115% minimum or HP rated or fuse holder size	Minimum size of starter	Controller termination temperature rating 60°C TW	60°C THW	75°C TW	75°C THW	Min. size of copper wire Wire size (AWG or kcmil)	Conduit (inches)
20	54	60*	60*	100	3	•	•	•	•	4	1
25	68	80	75	100	3	•	•			3	1 1/4
								•	•	3	1
30	80	100	90	100	3	•	•			1	1 1/4
								•	•	3	1 1/4
40	104	125	110	200	4	•	•			2/0	1 1/2
								•	•	1	1 1/4
50	130	150	150	200	4	•	•			3/0	2
								•	•	2/0	1 1/2

* Fuse reducers required.

6-38

3φ 230 V MOTORS AND CIRCUITS – 240 V SYSTEM (cont.)

Size of motor HP	Amp	Motor overload protection Low-peak or Fusetron®		Switch 115% minimum or HP rated or fuse holder size	Minimum size of starter	Controller termination temperature rating				Minimum size of copper wire and trade conduit	
		Motor less than 40°C or greater than 1.15 SF (Max. fuse 125%)	All other motors (Max. fuse 115%)			60°C TW	THW	75°C TW	THW	Wire size (AWG or kcmil)	Conduit (inches)
75	192	225	200*	400	5	•	•			300	2 1/2
									•	250	2 1/2
100	248	300	250	400	5	•	•			500	3
								°	•	350	2 1/2
150	360	450	400*	600	6	•	•			300-2φ	2-2-1/2
									•	4/0-2φ	2-2

* Fuse reducers required.

6-39

3φ 460 V MOTORS AND CIRCUITS – 480 V SYSTEM

Size of motor HP	Amp	Motor overload protection Low-peak or Fusetron® — Motor less than 40°C or greater than 1.15 SF (Max. fuse 125%)	Motor overload protection Low-peak or Fusetron® — All other motors (Max. fuse 115%)	Switch 115% minimum or HP rated or fuse holder size	Minimum size of starter	Controller termination temperature rating — 60°C TW	Controller termination temperature rating — 60°C THW	Controller termination temperature rating — 75°C TW	Controller termination temperature rating — 75°C THW	Minimum size of copper wire and trade conduit — Wire size (AWG or kcmil)	Minimum size of copper wire and trade conduit — Conduit (inches)
1/2	1	1 1/4	1 1/8	30	00	•	•	•	•	14	1/2
3/4	1.4	1 6/10	1 6/10	30	00	•	•	•	•	14	1/2
1	1.8	2 1/4	2	30	00	•	•	•	•	14	1/2
1 1/2	2.6	3 2/10	2 6/10	30	00	•	•	•	•	14	1/2
2	3.4	4	3 1/2	30	00	•	•	•	•	14	1/2
3	4.8	5 6/10	5	30	0	•	•	•	•	14	1/2
5	7.6	9	8	30	0	•	•	•	•	14	1/2
7 1/2	11	12	12	30	1	•	•	•	•	14	1/2
10	14	17 1/2	15	30	1	•	•	•	•	14	1/2
15	21	25	20	30	2		•	•	•	10	1/2
20	27	30*	30*	60	2				•	8	3/4
										10	1/2

* Fuse reducers required.

3φ 460 V MOTORS AND CIRCUITS – 480 V SYSTEM (cont.)

Size of motor HP	Amp	Motor overload protection Low-peak or Fusetron® — Motor less than 40°C or greater than 1.15 SF (Max. fuse 125%)	All other motors (Max. fuse 115%)	Switch 115% minimum or HP rated or fuse holder size	Minimum size of starter	Controller termination temperature rating 60°C TW	60°C THW	75°C TW	75°C THW	Wire size (AWG or kcmil)	Conduit (inches)
25	34	40	35	60	2	•	•		•	6 / 8	1 / 3/4
30	40	50	45	60	3	•	•		•	6 / 8	1 / 3/4
40	52	60*	60*	100	3	•	•		•	4 / 6	1 / 1
50	65	80	70	100	3	•	•		•	3 / 4	1 1/4 / 1
60	77	90	80	100	4	•	•		•	1 / 3	1 1/4 / 1 1/4
75	96	110	110	200	4	•	•		•	1/0 / 1	1 1/2 / 1 1/4

* Fuse reducers required.

6-41

Size of motor		Motor overload protection Low-peak or Fusetron®		Switch 115% minimum or HP rated or fuse holder size	Minimum size of starter	Controller termination temperature rating				Minimum size of copper wire and trade conduit	
		Motor less than 40°C or greater than 1.15 SF (Max. fuse 125%)	All other motors (Max. fuse 115%)			60°C		75°C			
HP	Amp					TW	THW	TW	THW	Wire size (AWG or kcmil)	Conduit (inches)
100	124	150	125	200	4	•	•		•	3/0 2/0	2 1-1/2
125	156	175	175	200	5	•	•	•	•	4/0 3/0	2 2
150	180	225	200*	400	5	•	•		•	300 4/0	2-1/2 2
200	240	300	250	400	5	•	•	•	•	500 350	3 2-1/2
250	302	350	325	400	6	•	•	•	•	4/0-2φ 3/0-2φ	2-2 2-2
300	361	450	400*	600	6	•	•		•	300-2φ 4/0-2φ	2—1-1/2 2-2

* Fuse reducers required.

DC MOTORS AND CIRCUITS

Size of motor HP	Amp	Motor overload protection Low-peak or Fusetron® Motor less than 40°C or greater than 1.15 SF (Max. fuse 125%)	All other motors (Max. fuse 115%)	Switch 115% minimum or HP rated or fuse holder size	Minimum size of starter	Controller termination temperature rating 60°C TW THW	Controller termination temperature rating 75°C TW THW	Wire size (AWG or kcmil)	Conduit (inches)
90 V									
1/4	4.0	5	4 1/2	30	0	• •	• •	14	1/2
1/3	5.2	6 1/4	5 6/10	30	0	• •	• •	14	1/2
1/2	6.8	8	7.5	30	0	• •	• •	14	1/2
3/4	9.6	12	10	30	0	• •	• •	14	1/2
1	12.2	15	12	30	0	• •	• •	14	1/2
120 V									
1/4	3.1	3 1/2	3 1/2	30	0	• •	• •	14	1/2
1/3	4.1	5	4 1/2	30	0	• •	• •	14	1/2
1/2	5.4	6 1/4	6	30	0	• •	• •	14	1/2
3/4	7.6	9	8	30	0	• •	• •	14	1/2
1	9.5	10	10	30	0	• •	• •	14	1/2

* Fuse reducers required.

DC MOTORS AND CIRCUITS (cont.)

Size of motor HP	Amp	Motor overload protection Low-peak or Fusetron® Motor less than 40°C or greater than 1.15 SF (Max. fuse 125%)	All other motors (Max. fuse 115%)	Switch 115% minimum or HP rated or fuse holder size	Minimum size of starter	Controller termination temperature rating 60°C TW THW	Controller termination temperature rating 75°C TW THW	Minimum size of copper wire and trade conduit® Wire size (AWG or kcmil)	Conduit (inches)
120 V Continued									
1 1/2	13.2	15	15	30	1	• •	• •	14	1/2
2	17	20	17 1/2	30	1	• •	• •	12	1/2
5	40	50	45	60	2		•	6	3/4
								8	3/4
10	76	90	80	100	3	• •	•	2	1
								3	1
180 V									
1/4	2	2 1/2	2 1/4	30	0	• •	• •	14	1/2
1/3	2.6	3 2/10	2 8/10	30	0	• •	• •	14	1/2
1/2	3.4	4	3 1/2	30	0	• •	• •	14	1/2
3/4	4.8	6	5	30	0	• •	• •	14	1/2

® Fuse reducers required.

6-44

DC MOTORS AND CIRCUITS (cont.)

Size of motor HP	Amp	Motor overload protection Low-peak or Fusetron® Motor less than 40°C or greater than 1.15 SF (Max. fuse 125%)	All other motors (Max. fuse 115%)	Switch 115% minimum or HP rated or fuse holder size	Minimum size of starter	Controller termination temperature rating 60°C TW THW		75°C TW THW		Minimum size of copper wire and trade conduit Wire size (AWG or kcmil)	Conduit (inches)
180 V Continued											
1	6.1	7 1/2	7	30	0	•	•	•	•	14	1/2
1 1/2	8.3	10	9	30	1	•	•	•	•	14	1/2
2	10.8	12	12	30	1	•	•	•	•	14	1/2
3	16	20	17 1/2	30	1	•	•	•	•	12	1/2
5	27	30*	30*	60	1		•		•	8	1/2
									•	8	3/4

* Fuse reducers required.

CONTROL RATINGS

Size	Load (V)	Maximum HP — Normal duty 1φ	Normal duty 3φ	Plugging & jogging duty 1φ	Plugging & jogging duty 3φ	Cont. amps	Service limit amps	Tungsten & ballast type lamp amps 480 V max.	Resistance heating (kW) 1φ	Resistance heating (kW) 3φ	Transformer switching 20 times 1φ	20 times 3φ	20-40 times 1φ	20-40 times 3φ	Capacitor kVA switching rating 3φ kVAR
00	115	1/2	—	—	—	9	11	—	1.15	2.0	—	—	—	—	—
	200	—	1 1/2	—	—	9	11	—	2.0	3.46	—	—	—	—	—
	230	1	1 1/2	—	—	9	11	—	2.3	4.0	—	—	—	—	—
	380	—	1 1/2	—	—	9	11	—	—	6.5	—	—	—	—	—
	460	—	2	—	—	9	11	—	4.6	8.0	—	—	—	—	—
	575	—	2	—	—	9	11	—	5.8	10.0	—	—	—	—	—
0	115	1	3	1/2	1 1/2	18	21	20	2.3	4.0	0.6	—	0.3	—	—
	200	2	3	—	1 1/2	18	21	20	4.0	6.92	1.2	1.8	0.6	0.9	—
	230	—	3	1	1 1/2	18	21	20	4.6	8.0	—	2.1	—	1.0	—
	380	—	5	—	—	18	21	20	—	13.1	2.4	4.2	1.2	2.1	—
	460	—	5	—	2	18	21	20	9.2	15.9	3.0	5.2	1.5	2.6	—
	575	—	5	—	2	18	21	—	11.5	19.9	—	—	—	—	—
1	115	2	—	1	—	27	32	30	3.5	6.0	1.2	—	0.6	—	—
	200	3	7 1/2	2	3	27	32	30	6	10.4	2.4	3.6	1.2	1.8	—
	230	—	7 1/2	2	3	27	32	30	6.9	11.9	—	4.3	—	2.1	—
	380	—	10	—	5	27	32	30	—	19.7	—	—	—	—	—
	460	—	10	—	5	27	32	30	13.8	23.9	4.9	8.5	2.5	4.3	—
	575	—	10	—	5	27	32	—	17.3	29.8	6.2	11.0	3.1	5.3	—

CONTROL RATINGS (cont.)

Size	Load (V)	Maximum HP Normal duty 1φ	Normal duty 3φ	Plugging & jogging duty 1φ	Plugging & jogging duty 3φ	Cont. amps	Service limit amps	Tungsten & ballast type lamp amps 480 V max.	Resistance heating (kW) 1φ	Resistance heating (kW) 3φ	Transformer 20 times 1φ	20 times 3φ	20-40 times 1φ	20-40 times 3φ	Capacitor kVA switching ratings 3φ kVAR
1P	115	3	—	1 1/2	—	35	42	45	5.8	—	—	—	—	—	—
	230	5	—	3	—	35	42	45	11.5	—	—	—	—	—	—
1 3/4	115	—	—	—	—	40	40	45	5.8	9.9	1.6	—	0.8	—	—
	200	—	10	—	5	40	40	45	10	17.3	—	4.9	—	2.4	—
	230	—	10	—	5	40	40	45	11.5	19.9	3.2	5.75	1.6	2.8	—
	380	—	15	—	7 1/2	40	40	45	—	32.9	—	—	—	—	—
	460	—	15	—	7 1/2	40	40	45	23	39.8	6.6	11.2	3.3	5.7	—
	575	—	15	—	7 1/2	40	40	—	28.8	49.7	8.1	14.5	4.1	7.1	—
2	115	3	—	2	—	45	52	60	8.1	13.9	2.1	—	1.0	—	—
	200	—	10	—	7 1/2	45	52	60	14	24.2	—	6.3	—	3.1	—
	230	7 1/2	10	5	10	45	52	60	16.1	27.8	4.1	7.2	2.1	3.6	8
	380	—	15	—	15	45	52	60	—	46.0	—	—	—	—	—
	460	—	25	—	15	45	52	60	32.2	55.7	8.3	14	4.2	7.2	16
	575	—	25	—	15	45	52	—	40.3	69.6	10.0	18	5.2	8.9	20
2 1/2	115	5	—	—	—	60	65	75	10.4	17.9	3.1	—	1.5	—	—
	200	—	15	—	10	60	65	75	18	31.1	—	9.1	—	4.6	—
	230	10	20	—	15	60	65	75	20.7	35.8	6.1	10.6	3.1	5.3	17.5
	380	—	20	—	20	60	65	75	—	59.2	—	—	—	—	—
	460	—	30	—	20	60	65	75	41.4	71.6	12	21	6.1	10.6	34.5
	575	—	30	—	20	60	65	—	51.8	89.5	15	26.5	7.6	13.4	43.5

6-47

CONTROL RATINGS (cont.)

Size	Load (V)	Max HP Normal duty 1φ	Normal duty 3φ	Plugging & jogging duty 1φ	Plugging & jogging duty 3φ	Cont. amps	Service limit amps	Tungsten & ballast type lamp amps 480 V max.	Resistance heating (kW) 1φ	Resistance heating (kW) 3φ	Transformer 20 times 1φ	20 times 3φ	20-40 times 1φ	20-40 times 3φ	Capacitor kVA switching ratings 3φ kVAR
3	115	7 1/2	—	—	—	90	104	100	14.4	24.8	4.1	—	2.0	—	—
	200	15	25	—	15	90	104	100	25	43.3	8.1	12	4.1	6.1	—
	230	—	30	—	20	90	104	100	28.8	50.0	—	14	—	7.0	27
	380	—	50	—	30	90	104	100	—	82.2	—	28	—	14	—
	460	—	50	—	30	90	104	100	57.5	99.4	16	35	8.1	14	53
	575	—	50	—	30	90	104	—	71.9	124	20	18	10	18	67
3 1/2	115	—	—	—	—	115	125	150	18.4	31.8	—	—	—	—	—
	200	—	30	—	20	115	125	150	32	55.4	11	16	5.4	8	—
	230	—	60	—	25	115	125	150	36.8	63.7	—	18.5	—	9.5	33.5
	380	—	60	—	30	115	125	150	—	105	—	37.5	—	18.5	—
	460	—	75	—	40	115	125	150	73.6	127	21.5	47	11.0	23.5	66.5
	575	—	75	—	40	115	125	—	92	159	37	—	13.5	—	83.5
4	200	—	40	—	25	135	156	200	39	67.5	14	20	6.8	10	—
	230	—	50	—	30	135	156	200	44.9	77.6	—	23	—	12	40
	380	—	75	—	50	135	156	200	—	128	—	47	—	23	—
	460	—	100	—	60	135	156	200	89.7	155	27	59	14	23	80
	575	—	100	—	60	135	156	—	112	194	34	—	17	29	100

CONTROL RATINGS (cont.)

Size	Load (V)	Maximum HP Normal duty 1φ	Maximum HP Normal duty 3φ	Maximum HP Plugging & jogging duty 1φ	Maximum HP Plugging & jogging duty 3φ	Cont. amps 3φ	Service limit amps	Tungsten & ballast type lamp amps 480 V max.	Resistance heating (kW) 1φ	Resistance heating (kW) 3φ	Transformer switching 20 times 1φ	Transformer switching 20 times 3φ	Transformer switching 20–40 times 1φ	Transformer switching 20–40 times 3φ	Capacitor kVA switching ratings 3φ kVAR
4 1/2	200	—	50	—	30	210	225	250	53	91.7	—	30.5	—	15	—
	230	—	75	—	40	210	225	250	60.9	105	20.5	35	10.4	18	60
	380	—	100	—	75	210	225	250	—	174	—	—	—	—	—
	460	—	150	—	100	210	225	250	122	211	40.5	70.5	20.5	35	120
	575	—	150	—	100	210	225	—	152	264	51	88	25.5	44	150

STANDARD MOTOR SIZES

Classification	Size (HP)
Milli	1, 1.5, 2, 3, 5, 7.5, 10, 15, 25, 35
Fractional	1/20, 1/12, 1/8, 1/6, 1/4, 1/3, 1/2, 3/4
Full	1, 1-1/2, 2, 3, 5, 7-1/2, 10, 15, 20, 25, 30, 40, 50, 60, 75, 100, 125, 150, 200, 250, 300
Full—Special Order	350, 400, 450, 500, 600, 700, 800, 900, 1000, 1250, 1500, 1750, 2000, 2250, 2500, 3000, 3500, 4000, 4500, 5000, 5500, 6000, 7000, 8000, 9000, 10,000, 11,000, 12,000, 13,000, 14,000, 15,000, 16,000, 17,000, 18,000, 19,000, 20,000, 22,500, 30,000, 32,500, 35,000, 37,500, 40,000, 45,000, 50,000

FULL-LOAD CURRENTS—DC MOTORS

Motor rating (HP)	Current (Amps)	
	120 V	240 V
1/4	3.1	1.6
1/3	4.1	2.0
1/2	5.4	2.7
3/4	7.6	3.8
1	9.5	4.7
1 1/2	13.2	6.6
2	17	8.5
3	25	12.2
5	40	20
7 1/2	48	29
10	76	38

FULL-LOAD CURRENTS—1ϕ, AC MOTORS

Motor rating (HP)	Current (Amps)	
	115 V	230 V
1/6	4.4	2.2
1/4	5.8	2.9
1/3	7.2	3.6
1/2	9.8	4.9
3/4	18.8	6.9
1	16	8
1 1/2	20	10
2	24	12
3	34	17
5	56	28
7 1/2	80	40
10	100	50

FULL-LOAD CURRENTS—3φ, AC INDUCTION MOTORS

Motor rating (HP)	Current (Amps)			
	208 V	230 V	460 V	575 V
1/4	1.11	.96	.48	.38
1/3	1.34	1.18	.59	.47
1/2	2.2	2.0	1.0	.8
3/4	3.1	2.8	1.4	1.1
1	4.0	3.6	1.8	1.4
1 1/2	5.7	5.2	2.6	2.1
2	7.5	6.8	3.4	2.7
3	10.6	9.6	4.8	3.9
5	16.7	15.2	7.6	6.1
7 1/2	24.0	22.0	11.0	9.0
10	31.0	28.0	14.0	11.0
15	46.0	42.0	21.0	17.0
20	59	54	27	22
25	75	68	34	27
30	88	80	40	32
40	114	104	52	41
50	143	130	65	52
60	169	154	77	62
75	211	192	96	77
100	273	248	124	99
125	343	312	156	125
150	396	360	180	144
200	—	480	240	192
250	—	602	301	242
300	—	—	362	288
350	—	—	413	337
400	—	—	477	382
500	—	—	590	472

STARTING METHODS: SQUIRREL-CAGE INDUCTION MOTORS

Starter Type	% Full-Voltage Value		
	Voltage at Motor	Line Current	Motor Output Torque
Full Voltage	100	100	100
Autotransformer			
80 pc tap	80	64*	64
65 pc tap	65	42*	42
50 pc tap	50	25*	25
Primary-reactor			
80 pc tap	80	80	64
65 pc tap	65	65	42
50 pc tap	50	50	25
Primary-resistor			
Typical rating	80	80	64
Part-winding			
Low-speed motors (1/2-1/2)	100	50	50
High-speed motors (1/2-1/2)	100	70	50
High-speed motors (2/3-1/3)	100	65	42
Wye Start-Delta Run (1/3-1/3)	100	33	33

*Autotransformer magnetizing current not included.
Magnetizing current usually less than 25 percent motor full-load current.

NEMA RATINGS OF 60HZ AC CONTACTORS IN AMPERES

Size	8 Hr. Open Rating (A)	3φ			1φ	
		200 V	230 V	230/460 V	115 V	230V
00	9	1 1/2	1 1/2	2	1/3	1
0	18	3	3	5	1	2
1	27	7 1/2	7 1/2	10	2	3
2	45	10	15	25	3	7 1/2
3	90	25	30	50	–	–
4	135	40	50	100	–	–
5	270	75	100	200	–	–
6	540	150	200	400	–	–
7	810	–	300	600	–	–
8	1215	–	450	900	–	–
9	2250	–	800	1600	–	–

SETTING BRANCH CIRCUIT PROTECTIVE DEVICES

Type of Motor	Percent of Full Load Current			
	Nontime Delay Fuse	Dual-Element (Time-Delay) Fuse	Instant. Trip Type Breaker	Time-Limit Breaker
All AC single-phase and polyphase squirrel-cage and synchronous motors with full-voltage, resistance or reactor starting:				
No code letter	300	175	700	250
Code letter F to V	300	175	700	250
Code letter B to E	250	175	700	200
Code letter A	150	150	700	150
All AC squirrel-cage and synchronous motors with auto-transformer starting:				
Code letter F to V	250	175	700	200
Code letter B to E	200	175	700	200
Code letter A	150	150	700	150
Wound Rotor	150	150	700	150
Direct Current				
Not more than 50 HP	150	150	250	150
More than 50 HP	150	150	175	150

MAXIMUM OCPD

$$OCPD = FLC \times R_M$$

where
R_M = maximum rating of OCPD

and
FLC = full load current (from motor nameplate or
NEC® Table 430.150)

Motor Type	Code Letter	FLC %				
		Motor Size	TDF	NTDF	ITB	ITCB
AC*	—	—	175	300	150	700
AC*	A	—	150	150	150	700
AC*	B—E	—	175	250	200	700
AC*	F—V	—	175	300	250	700
DC	—	1/8 To 50 HP	150	150	150	250
DC	—	Over 50 HP	150	150	150	175

*full-voltage and resistor starting

STANDARD SIZES OF FUSES AND CBS

NEC® 240.6 lists standard amperage ratings of fuses and
fixed-trip circuit breakers as follows:

15	20	25	30	35	40
45	50	60	70	80	90
100	110	125	150	175	200
225	250	300	350	400	450
500	600	700	800	1000	1200
1600	2000	2500	3000	4000	5000	6000

LOCKED ROTOR CURRENT

Apparent, 1ϕ	Apparent, 3ϕ	True, 1ϕ	True, 3ϕ
$LRC = \dfrac{1000 \times HP \times kVA/HP}{V}$	$LRC = \dfrac{1000 \times HP \times kVA/HP}{V \times \sqrt{3}}$	$LRC = \dfrac{1000 \times HP \times kVA/HP}{V \times PF \times E_{ff}}$	$LRC = \dfrac{1000 \times HP \times kVA/HP}{V \times \sqrt{3} \times PF \times E_{ff}}$
where	where	where	where
LRC = locked rotor current (in amps)	LRC = locked rotor current (in amps)	LRC = locked rotor current (in amps)	LRC = locked rotor current (in amps)
1000 = multiplier for kilo	1000 = multiplier for kilo	1000 = multiplier for kilo	1000 = multiplier for kilo
HP = horsepower	HP = horsepower	HP = horsepower	HP = horsepower
kVA/HP = kilovolt amps per horsepower	kVA/HP = kilovolt amps per horsepower	kVA/HP = kilovolt amps per horsepower	kVA/HP = kilovolt amps per horsepower
V = volts	V = volts	V = volts	V = volts
	$\sqrt{3}$ = 1.732	PF = power factor	$\sqrt{3}$ = 1.732
		E_{ff} = motor efficiency	

MOTOR POWER FORMULAS — COST SAVINGS

Power Consumed	Operating Cost	Annual Savings
$$P = \frac{HP \times 746}{E_{ff}}$$ where P = power consumed (W) HP = horsepower 746 = constant E_{ff} = efficiency (%)	$$C_{/hr} = \frac{P_{/hr} \times C_{/kWh}}{1000}$$ where $C_{/hr}$ = operating cost per hour $P_{/hr}$ = power consumed per hour $C_{/kWh}$ = cost per kilowatt hour 1000 = constant to remove kilo	$$S_{Ann} = C_{Ann\ Std} - C_{Ann\ Eff}$$ where S_{Ann} = annual cost savings $C_{Ann\ Std}$ = annual operating cost for standard motor $C_{Ann\ Eff}$ = annual operating cost for energy-efficient motor

TYPICAL MOTOR POWER FACTORS

HP	Speed (rpm)	Power Factor at		
		1/2 load	3/4 load	full load
0—5	1800	.72	.82	.84
5.01—20	1800	.74	.84	.86
20.1—100	1800	.79	.86	.89
100.1—300	1800	.81	.88	.91

EFFICIENCY FORMULAS

Input and Output Power Known	Horsepower and Power Loss Known
$$E_{ff} = \frac{P_{out}}{P_{in}}$$ where E_{ff} = efficiency (%) P_{out} = output power (W) P_{in} = input power (W)	$$E_{ff} = \frac{746 \times HP}{746 \times HP + W_l}$$ where E_{ff} = efficiency (%) 746 = constant HP = horsepower W_l = watts lost

VOLTAGE UNBALANCE FORMULA

$$V_u = \frac{V_d}{V_a} \times 100$$

where
V_u = voltage unbalance (%)
V_d = voltage deviation (V)
V_a = voltage average (V)
100 = constant

VOLTAGE VARIATION CHARACTERISTICS

Performance Characteristics	10% above Rated Voltage	10% below Rated Voltage
Starting current	+10% to +12%	−10% to −12%
Full-load current	-7%	+11%
Motor torque	+20% to +25%	−20% to −25%
Motor efficiency	Little change	Little change
Speed	+1%	-1.5%
Temperature rise	−3°C to −4°C	+6°C to +7°C

FREQUENCY VARIATION CHARACTERISTICS

Performance Characteristics	5% above Rated Frequency	5% below Rated Frequency
Starting current	−5% to −6%	+5% to +6%
Full-load current	−1%	+1%
Motor torque	−10%	+11%
Motor efficiency	Slight increase	Slight decrease
Speed	+5%	−5%
Temperature rise	Slight decrease	Slight increase

GENERAL EFFECT OF VOLTAGE VARIATION ON DIRECT-CURRECT MOTOR CHARACTERISTICS

□ =INCREASE   ● =DECREASE

Voltage Variation	Starting and Max. Run Torque	Full-load Speed	EFFICIENCY			Full-load Current	Temperature Rise, Full Load	Maximum Overload Capacity	Magnetic Noise
			Full Load	3/4 Load	1/2 Load				
SHUNT-WOUND									
120% Voltage	□ 30%	□ 110%	Slight □	No Change	Slight □	● 17%	Main field □ Commutating field and armature ●	□ 30%	Slight □
110% Voltage	□ 15%	□ 105%	Slight □	No Change	Slight □	● 8.5%	Main field □ Commutating field and armature ●	□ 15%	Slight □
90% Voltage	● 16%	● 95%	Slight ●	No Change	Slight □	□ 11.5%	Main field ● Commutating field and armature □	● 16%	Slight ●
COMPOUND-WOUND									
120% Voltage	□ 30%	□ 112%	Slight □	No Change	Slight □	● 17%	Main field □ Commutating field and armature ●	□ 30%	Slight □
110% Voltage	□ 15%	□ 106%	Slight □	No Change	Slight □	● 8.5%	Main field □ Commutating field and armature ●	□ 15%	Slight □
90% Voltage	● 16%	● 94%	Slight ●	No Change	Slight □	□ 11.5%	Main field ● Commutating field and armature □	● 16%	Slight ●

NOTES:—Starting current is controlled by starting resistor.
This table shows general effects, which will vary somewhat for specific ratings.

GENERAL EFFECT OF VOLTAGE AND FREQUENCY VARIATION ON INDUCTION-MOTOR CHARACTERISTICS

NOTE: This table shows general effects, which will vary somewhat for specific ratings. □ =INCREASE   ● =DECREASE

		Starting and Max. Running Torque	Synchronous Speed	% Slip	Full-load Speed	EFFICIENCY			POWER-FACTOR			Full-load Current	Starting Current	Temperature Rise, Full Load	Maximum Overload Capacity	Magnetic Noise, No Load in Particular
						Full Load	3/4 Load	1/2 Load	Full Load	3/4 Load	1/2 Load					
Voltage Variation	120% Voltage	□ 44%	No Change	● 30%	□ 1.5%	Small □	1/2 to 2 points ●	7 to 20 points ●	5 to 15 points ●	10 to 30 points ●	15 to 40 points ●	11% ●	□ 25%	● 5 to 6C	□ 44%	Noticeable □
	110% Voltage	□ 21%	No Change	● 17%	□ 1%	□ 1/2 to 1 point	Practically no change ●	1 to 2 points ●	3 points ●	4 points ●	5 to 6 points ●	● 7%	□ 10 to 12%	● 3 to 4C	□ 21%	Slight □
	Function of Voltage	(Voltage)² □	Constant	$\frac{1}{(\text{Voltage})^2}$ ●	(Syn speed – slip)	—	—	—	—	—	—	—	Voltage □	—	(Voltage)² ●	—
	90% Voltage	● 19%	No Change	□ 23%	● 1½%	● 2 points	Practically no change ●	□ 1 to 2 points	□ 1 point	□ 2 to 3 points	□ 4 to 5 points	□ 11%	● 10 to 12%	□ 6 to 7C	● 19%	Slight ●
Frequency Variation	105% Frequency	● 10%	□ 5%	Practically no change	□ 5%	Slight □	Slight □	Slight □	Slight □	Slight □	Slight □	Slight ●	● 5 to 6%	Slight ●	Slight ●	Slight ●
	Function of Frequency	$\frac{1}{(\text{Frequency})^2}$ ●	Frequency □	Practically no change	(Syn speed – slip)	—	—	—	—	—	—	—	$\frac{1}{(\text{Frequency})}$ ●	—	—	—
	95% Frequency	□ 11%	● 5%	Practically no change	● 5%	Slight ●	Slight ●	Slight ●	Slight ●	Slight ●	Slight ●	Slight □	□ 5 to 6%	Slight □	—	Slight □

HEATER TRIP CHARACTERISTICS

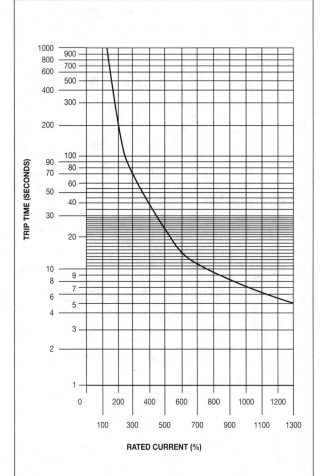

HEATER SELECTIONS

Heater number	Full-load current (A)*				
	Size 0	Size 1	Size 2	Size 3	Size 4
10	.20	.20	—	—	—
11	.22	.22	—	—	—
12	.24	.24	—	—	—
13	.27	.27	—	—	—
14	.30	.30	—	—	—
15	.34	.34	—	—	—
16	.37	.37	—	—	—
17	.41	.41	—	—	—
18	.45	.45	—	—	—
19	.49	.49	—	—	—
20	.54	.54	—	—	—
21	.59	.59	—	—	—
22	.65	.65	—	—	—
23	.71	.71	—	—	—
24	.78	.78	—	—	—
25	.85	.85	—	—	—
26	.93	.93	—	—	—
27	1.02	1.02	—	—	—
28	1.12	1.12	—	—	—
29	1.22	1.22	—	—	—
30	1.34	1.34	—	—	—
31	1.48	1.48	—	—	—
32	1.62	1.62	—	—	—
33	1.78	1.78	—	—	—
34	1.96	1.96	—	—	—
35	2.15	2.15	—	—	—
36	2.37	2.37	—	—	—
37	2.60	2.60	—	—	—

*Full-load current (A) does not include FLC x 1.15 or 1.25

HEATER SELECTIONS (cont.)

Heater number	Full-load current (A)*				
	Size 0	Size 1	Size 2	Size 3	Size 4
38	2.86	2.86	—	—	—
39	3.14	3.14	—	—	—
40	3.45	3.45	—	—	—
41	3.79	3.79	—	—	—
42	4.17	4.17	—	—	—
43	4.58	4.58	—	—	—
44	5.03	5.03	—	—	—
45	5.53	5.53	—	—	—
46	6.08	6.08	—	—	—
47	6.68	6.68	—	—	—
48	7.21	7.21	—	—	—
49	7.81	7.81	7.89	—	—
50	8.46	8.46	8.57	—	—
51	9.35	9.35	9.32	—	—
52	10.00	10.00	10.1	—	—
53	10.7	10.7	11.0	12.2	—
54	11.7	11.7	12.0	13.3	—
55	12.6	12.6	12.9	14.3	—
56	13.9	13.9	14.1	15.6	—
57	15.1	15.1	15.5	17.2	—
58	16.5	16.5	16.9	18.7	—
59	18.0	18.0	18.5	20.5	—
60	—	19.2	20.3	22.5	23.8
61	—	20.4	21.8	24.3	25.7
62	—	21.7	23.5	26.2	27.8
63	—	23.1	25.3	28.3	30.0
64	—	24.6	27.2	30.5	32.5
65	—	26.2	29.3	33.0	35.0

*Full-load current (A) does not include FLC x 1.15 or 1.25

HEATER SELECTIONS (cont.)

Heater number	Full-load current (A)*				
	Size 0	Size 1	Size 2	Size 3	Size 4
66	—	27.8	31.5	36.0	38.0
67	—	—	33.5	39.0	41.0
68	—	—	36.0	42.0	44.5
69	—	—	38.5	38.5	48.5
70	—	—	41.0	49.5	52
71	—	—	43.0	43.0	57
72	—	—	46.0	58	61
73	—	—	—	63	67
74	—	—	—	68	72
75	—	—	—	73	77
76	—	—	—	78	84
77	—	—	—	83	91
78	—	—	—	88	97
79	—	—	—	—	103
80	—	—	—	—	111
81	—	—	—	—	119
82	—	—	—	—	127
83	—	—	—	—	133

*Full-load current (A) does not include FLC x 1.15 or 1.25

HEATER AMBIENT TEMPERATURE CORRECTION

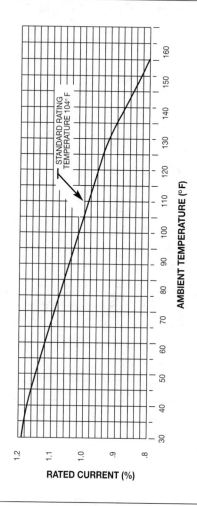

STANDARD RATING
TEMPERATURE 104° F

RATED CURRENT (%)

AMBIENT TEMPERATURE (°F)

HEATING ELEMENT SPECIFICATIONS

Material State	Heated Material	Outer Cover Material	Maximum Watt Density	Material Operating Temperature
Liquid	Gasoline	Iron/steel	18-22	300
	Oil (low viscosity)	Steel/copper	20-25	Up to 180
	Oil (medium viscosity)	Steel/copper	12-18	Up to 180
	Oil (high viscosity)	Steel/copper	5-7	Up to 180
	Vegetable oil (cooking)	Copper	28-32	400
	Water (washroom)	Copper	70-90	140
	Water (process)	Copper	40-50	212
Air/Gas	Still air (ovens)	Steel/stainless steel	28-32 18-22 8-10 2-3	Up to 700 Up to 1000 Up to 1200 Up to 1500
	Air (moving at 10 fps)	Aluminum steel	30-34 23-26 14-16 2-3	Up to 700 Up to 1000 Up to 1200 Up to 1500

HEATING ELEMENT SPECIFICATIONS (cont.)

Material State	Heated Material	Outer Cover Material	Maximum Watt Density	Material Operating Temperature
Solid	Asphalt	Iron/steel	9-12 8-10 6-8 5-6	Up to 200 Up to 300 Up to 400 Up to 500
	Molten tin (heated in pot)	Iron/steel	20-22	600
	Steel tubing (heated indirectly)	Iron/steel	45-48 50-52 54-56	500 750 1000

DIRECT CURRENT MOTOR TROUBLESHOOTING GUIDE

Problem	Possible Cause	Corrective Action
Motor will not start	Blown fuse or open CB	Test the OCPD. If voltage is present at the input, but not the output of the OCPD, the fuse is blown or the CB is open. Check the rating of the OCPD. It should be at least 125% of the motor's FLC.
	Motor overload on starter tripped	Allow overloads to cool. Reset overloads. If reset overloads do not start motor, test the starter.
	No brush contact	Check brushes. Replace, if worn.
	Open control circuit between incoming power and motor	Check for cleanliness, tightness, and breaks. Use a voltmeter to test the circuit starting with the incoming power and moving to the motor terminals. Voltage generally stops at the problem area.
Fuse, CB, or overloads retrip after service	Excessive load	If the motor is loaded to excess or is jammed, the circuit OCPD will open. Disconnect the load from the motor. If the motor now runs properly, check the load. If the motor does not run and the fuse or CB opens, the problem is with the motor or control circuit. Remove the motor from the control circuit and connect it directly to the power source. If the motor runs properly, the problem is in the control circuit. Check the control circuit. If the motor opens the fuse or CB again, the problem is in the motor. Replace or service the motor.
	Motor shaft does not turn	Disconnect the motor from the load. If the motor shaft still does not turn, the bearings are frozen. Replace or service the motor.
Brushes chip or break	Brush material is too weak or the wrong type for motor's duty rating	Replace with better grade or type of brush. Consult manufacturer if problem continues.
	Brush face is overheating and losing brush bonding material	Check for an overload on the motor. Reduce the load as required. Adjust brush holder arms.

DIRECT CURRENT MOTOR TROUBLESHOOTING GUIDE (cont.)

Problem	Possible Cause	Corrective Action
Brushes chip or break	Brush holder is too far from commutator	Too much space between the brush holder and the surface of the commutator allows the brush end to chip or break. Set correct space between brush holder and commutator.
	Brush tension is incorrect	Adjust brush tension so the brush rides freely on the commutator.
Brushes spark	Worn brushes	Replace worn brushes. Service the motor if rapid brush wear, excessive sparking, chipping, breaking, or chattering is present.
	Commutator is concentric	Grind commutator and undercut mica. Replace commutator if necessary.
	Excessive vibration	Balance armature. Check brushes. They should be riding freely.
Rapid brush wear	Wrong brush material, type, or grade	Replace with brushes recommended by manufacturer.
	Incorrect brush tension	Adjust brush tension so the brush rides freely on the commutator.
Motor overheats	Improper ventilation	Clean all ventilation openings. Vacuum or blow dirt out of motor with low-pressure, dry, compressed air.
	Motor is overloaded	Check the load for binding. Check shaft straightness. Measure motor current under operating conditions. If the current is above the listed current rating, remove the motor. Remeasure the current under no-load conditions. if the current is excessive under load but not when unloaded, check the load. If the motor draws excessive current when disconnected, replace or service the motor.

SHADED POLE MOTOR TROUBLESHOOTING GUIDE

Problem	Possible Cause	Corrective Action
Motor will not start	Blown fuse or open CB	Test OCPD. If voltage is present at the input, but not the output of the OCPD, the fuse is blown or the CB is open. Check the rating of the OCPD. It should be at least 125% of the motor's FLC.
	Motor overload on starter tripped	Allow overloads to cool. Reset overloads. If reset overloads do not start the motor, test the starter.
	Low or no voltage applied to motor	Check the voltage at the motor terminals. The voltage must be present and within 10% of the motor nameplate voltage. If voltage is present at the motor but the motor is not operating, remove the motor from the load the motor is driving. Reapply power to the motor. If the motor runs, the problem is with the load. If the motor does not run, the problem is with the motor. Replace or service the motor.
	Open control circuit between incoming power and motor	Check for cleanliness, tightness, and breaks. Use a voltmeter to test the circuit starting with the incoming power and moving to the motor terminals. Voltage generally stops at the problem area.
Fuse, CB, or overloads retrip after service	Excessive load	If the motor is loaded to excess or jammed, the circuit OCPD will open. Disconnect the load from the motor. If the motor now runs properly, check the load. If the motor does not run and the fuse or CB opens, the problem is with the motor or control circuit. Remove the motor from the control circuit and connect it directly to the power source. If the motor runs properly, the problem is in the control circuit. Check the control circuit, if the motor opens the fuse or CB again, the problem is in the motor. Replace or service the motor.
Excessive noise	Unbalanced motor or load	An unbalanced motor or load causes vibration, which causes noise. Realign the motor and load. Check for excessive end play or loose parts. If the shaft is bent, replace the rotor or motor.
	Dry or worn bearings	Dry or worn bearings cause noise. Bearings may be dry due to dirty oil, oil not reaching the shaft, or motor overheating. Oil bearings as recommended. If noise remains, replace the bearings or motor.
	Excessive grease	Ball bearings that have excessive grease may cause bearings to overheat. Overheated bearings cause noise. Remove excess grease.

SPLIT-PHASE MOTOR TROUBLESHOOTING GUIDE

Problem	Possible Cause	Corrective Action
Motor will not start	Thermal cutout switch is open	Reset the thermal switch. Caution: Resetting the thermal switch may automatically start the motor.
	Blown fuse or open CB	Test the OCPD, if voltage is present at the input, but not the output of the OCPD, the fuse is blown or the CB is open. Check the rating of the OCPD. It should be at least 125% of the motor's FLC.
	Motor overload on starter tripped	Allow overloads to cool. Reset overloads. If reset overloads do not start the motor, test the starter.
	Low or no voltage applied to motor	Check the voltage at the motor terminals. The voltage must be present and within 10% of the motor nameplate voltage. If voltage is present at the motor but the motor is not operating, remove the motor from the load the motor is driving. Reapply power to the motor. If the motor runs, the problem is with the load. if the motor does not run, the problem is with the motor. Replace or service the motor.
	Open control circuit between incoming power and motor	Check for cleanliness, tightness, and breaks. Use a voltmeter to test the circuit starting with the incoming power and moving to the motor terminals. Voltage generally stops at the problem area.
	Starting winding not receiving power	Check the centrifugal switch to make sure it connects the starting winding when the motor is OFF.
Fuse, CB, or overloads retrip after service	Blown fuse or open CB	Test the OCPD. If voltage is present at the input, but not the output of the OCPD, the fuse is blown or the CB is open. Check the rating of the OCPD. It should be at least 125% of the motor's FLC.
	Motor overload on starter tripped	Allow overloads to cool. Reset overloads. If reset overloads do not start the motor, test the starter.

SPLIT-PHASE MOTOR TROUBLESHOOTING GUIDE (cont.)

Problem	Possible Cause	Corrective Action
Fuse, CB, or overloads retrip after service	Low or no voltage applied to motor	Check the voltage at the motor terminals. The voltage must be present and within 10% of the motor nameplate voltage. If voltage is present at the motor but the motor is not operating, remove the motor from the load the motor is driving. Reapply power to the motor. If the motor runs, the problem is with the load. If the motor does not run, the problem is with the motor. Replace or service the motor.
	Open control circuit between incoming power and motor	Check for cleanliness, tightness, and breaks. Use a voltmeter to test the circuit starting with the incoming power and moving to the motor terminals. Voltage generally stops at the problem area.
	Motor shaft does not turn	Disconnect the motor from the load. If the motor shaft still does not turn, the bearings are frozen. Replace or service the motor.
Motor produces electric shock	Broken or disconnected ground strap	Connect or replace ground strap. Test for proper ground.
	Hot power lead at motor connecting terminals is touching motor frame	Disconnect the motor. Open the motor terminal box and check for poor connections, damaged insulation, or leads touching the frame. Service and test motor for ground.
	Motor winding shorted to frame	Remove, service, and test motor.
Motor overheats	Starting windings are not being removed from circuit as motor accelerates	When the motor is turned OFF, a distinct click should be heard as the centrifugal switch closes.
	Improper ventilation	Clean all ventilation openings. Vacuum or blow dirt out of motor with low-pressure, dry, compressed air.

SPLIT-PHASE MOTOR TROUBLESHOOTING GUIDE (cont.)

Problem	Possible Cause	Corrective Action
Motor overheats	Motor is overloaded	Check the load for binding. Check shaft straightness. Measure motor current under operating conditions. If current is above the listed current rating, remove the motor. Remeasure the current under no-load conditions. If the current is excessive under load but not when unloaded, check the load. If the motor draws excessive current when disconnected, replace or service the motor.
	Dry or worn bearings	Dry or worn bearings cause noise. The bearings may be dry due to dirty oil, oil not reaching the shaft, or motor overheating. Oil the bearings as recommended. If noise remains, replace the bearings or the motor.
	Dirty bearings	Clean or replace bearings.
Excessive noise	Excessive end-play	Check and play by trying to move the motor shaft in and out. Add end-play washers as required.
	Unbalanced motor or load	An unbalanced motor or load causes vibration, which causes noise. Realign the motor and load. Check for excessive end play or loose parts. If the shaft is bent, replace the rotor or motor.
	Dry or worn bearings	Dry or worn bearings cause noise. The bearings may be dry due to dirty oil, oil not reaching the shaft, or motor overheating. Oil the bearings as recommended. If noise remains, replace the bearings or the motor.
	Excessive grease	Ball bearings that have excessive grease may cause the bearings to overheat. Overheated bearings cause noise. Remove any excess grease.

6-74

THREE-PHASE MOTOR TROUBLESHOOTING GUIDE

Problem	Possible Cause	Corrective Action
Motor will not start	Wrong motor connections	Most 3ϕ motors are dual-voltage. Check for proper motor connections.
	Blown fuse or open CB	Test the OCPD. If voltage is present at the input, but not the output of the OCPD, the fuse is blown or the CB is open. Check the rating of the OCPD. It should be at least 125% of the motor's FLC.
	Motor overload on starter tripped	Allow overloads to cool. Reset overloads. If reset overloads do not start the motor, test the starter.
	Low or no voltage applied to motor	Check the voltage at the motor terminals. The voltage must be present and within 10% of the motor nameplate voltage. If voltage is present at the motor but the motor is not operating, remove the motor from the load the motor is driving. Reapply power to the motor. If the motor runs, the problem is with the load. If the motor does not run, the problem is with the motor. Replace or service the motor.
	Open control circuit between incoming power and motor	Check for cleanliness, tightness, and breaks. Use a voltmeter to test the circuit starting with the incoming power and moving to the motor terminals. Voltage generally stops at the problem area.
Fuse, CB, or over-loads retrip after ser-vice	Power not applied to all three lines	Measure voltage at each power line. Correct any power supply problems.
	Blown fuse or open CB	Test the OCPD. If voltage is present at the input, but not the output of the OCPD, the fuse is blown or the CB is open. Check the rating of the OCPD. It should be at least 125% of the motor's FLC.
	Motor overload on starter tripped	Allow overloads to cool. Reset overloads. If reset overloads do not start the motor, test the starter.

THREE-PHASE MOTOR TROUBLESHOOTING GUIDE (cont.)

Problem	Possible Cause	Corrective Action
Fuse, CB, or overloads retrip after service	Low or no voltage applied to motor	Check the voltage at the motor terminals. The voltage must be present and within 10% of the motor nameplate voltage. If voltage is present at the motor but the motor is not operating, remove the motor from the load the motor is driving. Reapply power to the motor. If the motor runs, the problem is with the load. If the motor does not run, the problem is with the motor. Replace or service the motor.
	Open control circuit between incoming power and motor	Check for cleanliness, tightness, and breaks. Use a voltmeter to test the circuit starting with the incoming power and moving to the motor terminals. Voltage generally stops at the problem area.
	Motor shaft does not turn	Disconnect the motor from the load. If the motor shaft still does not turn, the bearings are frozen. Replace or service the motor.
Motor overheats	Motor is single phasing	Check each of the 3φ power lines for correct voltage.
	Improper ventilation	Clean all ventilation openings. Vacuum or blow dirt out of motor with low-pressure, dry, compressed air.
	Motor is overloaded	Check the load for binding. Check shaft straightness. Measure motor current under operating conditions. If the current is above the listed current rating, remove the motor. Remeasure the current under no-load conditions. If the current is excessive under load but not when unloaded, check the load. If the motor draws excessive current when disconnected, replace or service the motor.

CONTACTOR AND MOTOR STARTER TROUBLESHOOTING GUIDE

Problem	Possible Cause	Corrective Action
Humming noise	Magnet pole faces misaligned	Realign. Replace magnet assembly if realignment is not possible.
	Too low voltage at coil	Measure voltage at coil. Check voltage rating of coil. Correct any voltage that is 10% less than coil rating.
	Pole face obstructed by foreign object, dirt, or rust	Remove any foreign object and clean as necessary. Never file pole faces.
Loud buzz noise	Shading coil broken	Replace coil assembly.
Controller fails to drop out	Voltage to coil not being removed	Measure voltage at coil. Trace voltage from coil to supply looking for shorted switch or contact if voltage is present.
	Worn or rusted parts causing binding	Clean rusted parts, Replace worn parts.
	Contact poles sticking	Checking for burning or sticky substance on contacts. Replace burned contacts. Clean dirty contacts.
	Mechanical interlock binding	Check to ensure interlocking mechanism is free to move when power is OFF. Replace faulty interlock.
Controller fails to pull in	No coil voltage	Measure voltage at coil terminals. Trace voltage loss from coil to supply voltage if voltage is not present.
	Too low voltage	Measure voltage at coil terminals. Correct voltage level if voltage is less than 10% of rated coil voltage. Check for a voltage drop as large loads are energized.
	Coil open	Measure voltage at coil. Remove coil if voltage is present and correct but coil does not pull in. Measure coil resistance for open circuit. Replace if open.

CONTACTOR AND MOTOR STARTER TROUBLESHOOTING GUIDE (cont.)

Problem	Possible Cause	Corrective Action
Controller fails to pull in	Coil shorted	Shorted coil may show signs of burning. The fuse or breakers should trip if coil is shorted. Disconnect one side of coil and reset if tripped. Remove coil and check resistance for short if protection device does not trip. Replace shorted coil. Replace any coil that is burned.
	Mechanical obstruction	Remove any obstructions.
Contacts badly burned or welded	Too high inrush current	Measure inrush current. Check load for problem if higher-than-rated load current. Change to larger controller if load current is correct but excessive for controller.
	Too fast load cycling	Change to larger controller if load cycled ON and OFF repeatedly.
	Too large overcurrent protection device	Size overcurrent protection to load and controller.
	Short circuit	Check fuses or breakers. Clear any short circuit.
	Insufficient contact pressure	Check to ensure contacts are making good connection.
Nuisance tripping	Incorrect overload size	Check size of overload against rated load current. Size up if permissible per NEC®.
	Lack of temperature compensation	Correct setting of overload if controller and load are at different ambient temperatures
	Loose connections	Check for loose terminal connection.

FAULTY SOLENOID PROBLEMS

Problem	Possible Causes	Comments
Failure to operate when energized	Complete loss of power to solenoid	Normally caused by blown fuse or control circuit problem.
	Low voltage applied to the solenoid	Voltage should be at least 85% of solenoid's rated value.
	Burned out solenoid coil	Normally evident by pungent odor caused by burnt insulation.
	Shorted coil	Normally a fuse is blown and continues to blow when changed.
	Obstruction of plunger movement	Normally caused by a broken part, misalignment, or the presence of a foreign object.
	Excessive pressure on solenoid plunger	Normally caused by excessive system pressure in solenoid-operated valves.
Failure to operate spring-return solenoids when de-energized	Faulty control circuit	Normally a problem of the control circuit not disengaging the solenoid's hold or memory circuit.
	Obstruction of plunger movement	Normally caused by a broken part, misalignment, or the presence of a foreign object.
	Excessive pressure on solenoid plunger	Normally caused by excessive system pressure in solenoid-operated valves.
Failure to operate electrically-operated return solenoids when de-energized	Complete loss of power to solenoid	Normally caused by a blown fuse or control circuit problem.
	Low voltage applied to solenoid	Voltage should be at least 85% of solenoid's rated value.

FAULTY SOLENOID PROBLEMS (cont.)

Problem	Possible Causes	Comments
Failure to operate electrically operated return solenoids when de-energized	Burned out solenoid coil	Normally evident by pungent odor caused by burnt insulation.
	Obstruction of plunger movement	Normally caused by broken part, misalignment, or presence of a foreign object.
	Excessive pressure on solenoid plunger	Normally caused by excessive system pressure in solenoid-operated valves.
Noisy operation	Solenoid housing vibrates	Normally caused by loose mounting screws.
	Plunger pole pieces do not make flush contact	An air gap may be present causing the plunger to vibrate. These symptoms are normally caused by foreign matter.
Erratic operation	Low voltage applied to the solenoid	Voltage should be at least 85% of the solenoid's rated voltage.
	System pressure may be low or excessive	Solenoid size is inadequate for the application.
	Control circuit is not operating properly	Conditions on the solenoid have increased to the point where the solenoid cannot deliver the required force.

SEMI-ANNUAL MOTOR MAINTENANCE CHECKLIST

Step	Operation	Mechanic
1	Turn OFF and lock out all power to the motor and its control circuit.	
2	Clean motor exterior and all ventilation ducts.	
3	Check motor's wire raceway.	
4	Check and lubricate bearings as needed.	
5	Check drive mechanism.	
6	Check brushes and commutator.	
7	Check slip rings.	
8	Check motor terminations.	
9	Check capacitors.	
10	Check all mounting bolts.	
11	Check and record line-to-line resistance.	
12	Check and record megohmmeter resistance from L1 to ground.	
13	Check motor controls.	
14	Reconnect motor and control circuit power supplies.	
15	Check line-to-line voltage for balance and level.	
16	Check line current draw against nameplate rating.	
17	Check and record inboard and outboard bearing temperatures.	

ANNUAL MOTOR MAINTENANCE CHECKLIST

Motor File #:	Serial #:
Date Installed:	Motor Location:

MFR:	Type:	Frame:
HP:	Volts:	Amps:
RPM:	Date Serviced:	

Step	Operation	Mechanic
1	Turn OFF and lock out all power to the motor and its control circuit.	
2	Clean motor exterior and all ventilation ducts.	
3	Uncouple motor from load and disassemble.	
4	Clean inside of motor.	
5	Check centrifugal switch assemblies.	
6	Check rotors, armatures, and field windings.	
7	Check all peripheral equipment.	
8	Check and lubricate bearings as needed.	
9	Check brushes and commutator.	
10	Check slip rings.	
11	Reassemble motor and couple to load.	

ANNUAL MOTOR MAINTENANCE CHECKLIST (cont.)

Step	Operation	Mechanic
12	Flush old bearing lubricant and replace.	
13	Check motor's wire raceway.	
14	Check drive mechanism.	
15	Check motor terminations.	
16	Check capacitors.	
17	Check all mounting bolts.	
18	Check and record line-to-line resistance.	
19	Check and record megohmmeter resistance from L and/or T to ground.	
20	Check and record insulation polarization index.	
21	Check motor controls.	
22	Reconnect motor and control circuit power supplies.	
23	Check line-to-line voltage for balance and level.	
24	Check line current draw against nameplate rating.	
25	Check and record inboard and outboard bearing temperatures.	
26	Check for vibration abnormalities.	
27	Observe motor for proper operation.	
28	Note any repairs/maintenance on service record.	

MOTOR REPAIR AND SERVICE RECORD

Motor File #: _____   Serial #: _____
Date Installed: _____ Motor Location: _____

MFR: _____   Type: _____   Frame: _____
HP: _____   Volts: _____ Amps: _____
RPM: _____   Filter Sizes: _____

Date	Operation	Mechanic

6-84

CHAPTER 7
TRANSFORMERS

TRANSFORMER CONNECTIONS

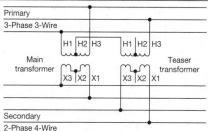

SCOTT CONNECTED THREE-PHASE TO TWO-PHASE

When two-phase power is required from a three-phase system, the Scott connection is used the most. The secondary may be three, four, or five wire. Special taps must be provided at 50% and 86.6% of normal primary voltage to make this connection.

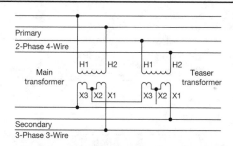

SCOTT CONNECTED TWO-PHASE TO THREE-PHASE

If it should be necessary to supply three-phase power from a two-phase system, the scott connection may be used again. The special taps must be provided on the secondary side, but the connection is similar to the three-phase to two-phase.

To obtain the Scott transformation without a special 86.6% tapped transformer, use one with 10% or two 5% taps to approximate the desired value. A small error of unbalance (overvoltage) occurs that requires care in application.

TRANSFORMER CONNECTIONS (cont.)

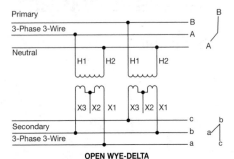

OPEN WYE-DELTA

When operating wye-delta and one phase is disabled, service may be maintained at reduced load as shown. The neutral in this case must be connected to the neutral of the setup bank through a copper conductor. The system is unbalanced, electro-statically and electro-magnetically, so that telephone interference may be expected if the neutral is connected to ground. The useful capacity of the open delta open wye bank is 87% of the capacity of the installed transformers when the two units are identical.

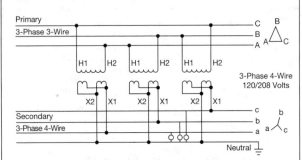

DELTA-WYE FOR LIGHTING AND POWER

In the previous banks the single-phase lighting load is all on one phase resulting in unbalanced primary currents in any one bank. To eliminate this difficulty, the delta-wye system finds many uses. Here the neutral of the secondary three-phase system is grounded and the single-phase loads are connected between the different phase wires and the neutral while the three-phase loads are connected to the phase wires. Thus, the single-phase load can be balanced on three phases in each bank and banks may be paralled if desired.

TRANSFORMER CONNECTIONS (cont.)

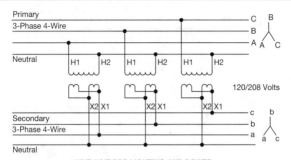

WYE-WYE FOR LIGHTING AND POWER

A system on which the primary voltage was increased from 2400 to 4160 V to increase the potential capacity. The previously delta connected distribution transformers are now connected from line to line. The secondaries are connected in wye. The primary neutral is connected to the neutral of the supply voltage through a metallic conductor and carried with the phase conductor to minimize telephone interference. If the neutral of the transformer is isolated from the system neutral an unstable condition results at the transformer neutral caused primarily by third harmonic voltages. If the transformer neutral is connected to ground, the possibility of telephone interference is enhanced and a possibility of resonance between the line capacitance to ground and the magnetizing impedance of the transformer.

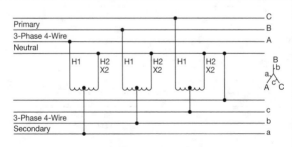

WYE-WYE AUTOTRANSFORMERS FOR SUPPLYING POWER FROM A THREE-PHASE FOUR-WIRE SYSTEM

When ratio of transformation from primary to secondary voltage is small, the best way of stepping down voltage is using autotransformers. It is necessary that the neutral of the autotransformer bank be connected to the system neutral.

TRANSFORMER CONNECTIONS (cont.)

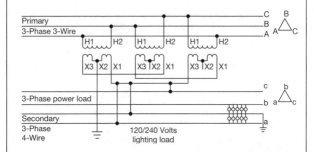

DELTA-DELTA FOR POWER AND LIGHTING

This connection is used to supply a small single-phase lighting load and three-phase power load simultaneously. As shown, the midtap of the secondary of one transformer is grounded. The small lighting load is connected across the transformer with the midtap and the ground wire common to both 120 V circuits. The single-phase lighting load reduces the available three-phase capacity. This requires special watt-hour metering.

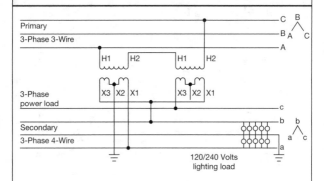

OPEN-DELTA FOR LIGHTING AND POWER

Where the secondary load is a combination of lighting and power, the open-delta connected bank is frequently used. This connection is used when the single-phase lighting load is large as compared with the power load. Here two different size transformers may be used with the lighting load connected across the larger rated unit.

TRANSFORMER CONNECTIONS (cont.)

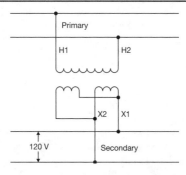

SINGLE-PHASE TO SUPPLY 120 V LIGHTING LOAD
The transformer is connected between high voltage line and load with the 120/240 V winding connected in parallel. This connection is used where the load is comparatively small and the length of the secondary circuit is short. It is often used for a single customer.

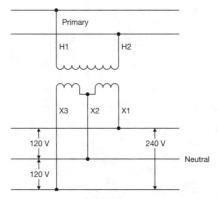

SINGLE-PHASE TO SUPPLY 120/240 V 3-WIRE LIGHTING AND POWER LOAD
Here the 120/240 V winding is connected in series and the mid-point brought out, making it possible to serve both 120 and 240 V loads simultaneously. This connection is used in most urban distribution circuits.

TRANSFORMER CONNECTIONS (cont.)

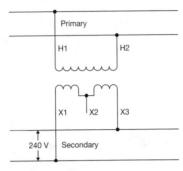

SINGLE-PHASE FOR POWER

In this case the 120/240 V winding is connected in series serving 240 V on a two-wire system. This connection is used for small industrial applications.

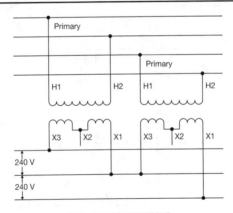

TWO-PHASE CONNECTIONS

This connection consists merely of two single-phase transformers operated 90° out of phase. For a three-wire secondary as shown, the common wire must carry 32 times the load current. In some cases, a four-wire or a five-wire secondary may be used.

TRANSFORMER SOUND LEVELS

kVA Rating	Sound Level (in dB)	Hearing Intensity	Example of Loudness
0–5	40	Barely Audible	Refrigerator running
6–9	40		
10–25	45		
26–50	45		
51–150	50	Quiet	Whisper
151–225	55		
226–300	55		
301–500	60	Moderate	Normal conversation

TRANSFORMING WINDING RATIOS

Tap Winding	Primary Voltage	Secondary Voltage	Ratio
Full Winding	2400	120/240	20:1/10:1
$4^1/2$%	2292	120/240	19:1/9.5:1
9%	2184	120/240	18:1/9:1
$13^1/2$%	2076	120/240	17:1/8.5:1

SATISFACTORY VOLTAGE LEVELS

Devices	Tolerable Under Voltage	Satisfactory Voltage	Tolerable Over Voltage
Lights, Motors, etc.	112	117	122–127
Stoves, Dryers, etc.	230	235	245–250

SINGLE-PHASE AC MOTORS
FULL-LOAD AMPERES

HP	115 V	208 V	230 V	MIN. TRANSFORMER kVA
1/6	4.4	2.4	2.2	.53
1/4	5.8	3.2	2.9	.70
1/3	7.2	4.0	3.6	.87
1/2	9.8	5.4	4.9	1.18
3/4	13.8	7.6	6.9	1.66
1	16	8.8	8	1.92
1-1/2	20	11	10	2.4
2	24	13.2	12	2.88
3	34	18.7	17	4.1
5	56	30.8	28	6.72
7-1/2	80	44	40	9.6
10	100	55	50	12

THREE-PHASE AC MOTORS
FULL-LOAD AMPERES

HP	208 V	230 V	460 V	575 V	MIN. TRANSFORMER kVA
1/2	2.2	2.0	1.0	0.8	0.9
3/4	3.1	2.8	1.4	1.1	1.2
1	4.0	3.6	1.8	1.4	1.5
1-1/2	5.7	5.2	2.6	2.1	2.1
2	7.5	6.8	3.4	2.7	2.7
3	10.6	9.6	4.8	3.9	3.8
5	16.7	15.2	7.6	6.1	6.3
7-1/2	24.2	22	11	9	9.2
10	31.8	28	14	11	11.2
15	46.2	42	21	17	16.6
20	59.4	54	27	22	21.6
25	74.8	68	34	27	26.6
30	88	80	40	32	32.4
40	114.4	104	52	41	43.2
50	143	130	65	52	52
60	169.4	154	77	62	64
75	211.2	192	96	77	80
100	272.8	248	124	99	103
125	343.2	312	156	125	130
150	396	360	180	144	150
200	528	480	240	192	200

SINGLE-PHASE TRANSFORMERS

kVA RATING	AMPERES			
	120 V	240 V	480 V	600 V
1	8.33	4.17	2.08	1.67
1-1/2	12.5	6.25	3.13	2.50
2	16.7	8.33	4.17	3.33
3	25.0	12.5	6.25	5.00
5	41.7	20.8	10.4	8.33
7-1/2	62.5	31.3	15.6	12.5
10	83.3	41.7	20.8	16.7
15	125	62.5	31.3	25.0
20	167	83.3	41.7	33.3
25	208	104	52.1	41.7
30	250	125	62.5	50
37-1/2	313	156	78.0	62.5
50	417	208	104	83.3
75	625	313	156	125
100	833	417	208	167
150	1250	625	313	250
167	1392	696	348	278
200	1667	833	417	333
250	2083	1042	521	417
333	2775	1388	694	555
500	4167	2083	1042	833

THREE-PHASE TRANSFORMERS

kVA RATING	AMPERES			
	208 V	240 V	480 V	600 V
3	8.3	7.2	3.6	2.9
6	16.6	14.4	7.2	5.8
9	25.0	21.6	10.8	8.7
15	41.6	36	18	14.4
20	55.6	48.2	24.1	19.3
25	69.5	60.2	30.1	24.1
30	83.0	72	36	28.8
37-1/2	104	90.3	45.2	36.1
45	125	108	54	43
50	139	120	60.2	48.2
60	167	145	72.3	57.8
75	208	180	90	72
100	278	241	120	96.3
112.5	312	270	135	108
150	415	360	180	144
200	554	480	240	192
225	625	540	270	216
300	830	720	360	288
400	1110	960	480	384
500	1380	1200	600	480
750	2080	1800	900	720
1000	2780	2400	1200	960
1500	4150	3600	1800	1440
2000	5540	4800	2400	1920

RF COIL WINDING FORMULAS

Using the equations below, the inductance of air-core coil can be calculated to within 1% or 2%.

T = Number of turns   I = inductance in microhenrys
W, C and H are measured in inches

Example 1:   Single Layer Coil

$$I = \frac{H^2 T^2}{9H + 10W}$$

Example 2:   Multiple Layer Coil

$$I = \frac{0.8\,(H^2 T^2)}{6H + 9W + 10C}$$

Example 3:   Single Layer, Single Row Coil

$$I = \frac{H^2 T^2}{8H + 11C}$$

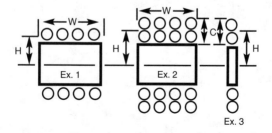

TURNS PER INCH vs. WIRE SIZE

Wire Gauge (AWG)	Turns Per Inch of Length		
	S.S.C.	Enamel	D.C.C.
40	194	282	89.7
39	181	248	86.6
38	166	224	83.6
37	154	198	80.3
36	143	175	77
35	132	158	73.5
34	120	143	70
33	110	127	66.3
32	101	113	62.6
31	92	101	59.2
30	83.3	90.5	55.5
29	74.8	81.6	51.8
28	68.6	72.7	48.5
27	61.5	64.9	45
26	55.6	58	41.8
25	50.4	51.7	38.6
24	45.3	46.3	35.6
23	40.6	41.3	31.6
22	36.5	37	30
21	32.7	33.1	26
20	29.4	29.4	23.8
19	26.4	26.4	21.8
18	23.6	23.6	19.8
17	21.2	21.2	18.1
16	18.9	18.9	16.4
15	-	16.8	14.7
14	-	15	13.8
13	-	13.5	12
12	-	12	10.9
11	-	10.7	9.8
10	-	9.6	8.9
9	-	8.6	7.8
8	-	7.6	7.1
7	-	-	6.2
6	-	-	5.6
5	-	-	5
4	-	-	4.5
3	-	-	4
2	-	-	3.6
1	-	-	3.3

Values will vary according to the manufacturer of the wire and enamel thickness.

COMMON TRANSFORMER kVA RATINGS

Single-Phase	Three-Phase
25	75
37.5	112.5
50	150
75	225
100	300
167	500
250	750
333	1000
500	1500
667	2000
833	2500
1000	3000
1250	3750
1667	5000
2000	6000
3333	7500
4000	10,000
5000	12,000
6667	15,000
8333	20,000
10,000	25,000
12.500	30,000
16,667	37,000
20,000	50,000
25,000	60,000
33,000	75,000

CHAPTER 8
ELECTRICAL PLAN SYMBOLS

To avoid confusion, ASA policy requires that the same symbol not be included in more than one Standard. If the same symbol were to be used in two or more Standards and one of these Standards were revised changing the meaning of the symbol, considerable confusion could arise over which symbol was correct, the revised or unrevised. The symbols in this category include, but are not limited to, those listed below.

GENERAL SYMBOLS

(M)	Electric motor
(G)	Electric generator
⧢	Power transformer
◁	Pothead (cable termination)
(WH)	Electric watthour meter

GENERAL SYMBOLS (*cont.*)

Symbol	Description
CB	Circuit element, e.g., circuit breaker
	Circuit breaker
	Fusible element
	Single-throw knife switch
	Double-throw knife switch
	Ground
	Battery

LIGHTING OUTLET SYMBOLS

CEILING   WALL

◯	—◯	Surface or pendant incandescent, mercury vapor, or similar lamp fixture
(R)	—(R)	Recessed incandescent, mercury vapor, or similar lamp fixture
[O]		Surface or pendant individual fluorescent fixture
[OR]		Recessed individual fluorescent fixture
[O |||]		Surface or pendant continuous-row fluorescent fixture
[OR |||]		Recessed continuous-row fluorescent fixture*
├——┼——┼——┤		Bare-lamp fluorescent strip**

* In the case of combination continuous-row fluorescent and incandescent spotlights, use combinations of the above Standard symbols.

** In the case of a continuous-row bare-lamp fluorescent strip above an area-wide diffusion means, show each fixture run, using the Standard symbol; indicate area of diffusing means and type of light shading and/or drawing notation.

LIGHTING OUTLET SYMBOLS (*cont.*)

CEILING	WALL	
(X)	—(X)	Surface or pendant exit light
(XR)	—(XR)	Recessed exit light
(B)	—(B)	Blanked outlet
(J)	—(J)	Junction box
(L)	—(L)	Outlet controlled by low-voltage switching when relay is installed in outlet box

RECEPTACLE OUTLET SYMBOLS

⊖	Single receptacle outlet
⊖=	Duplex receptacle outlet
⊖≡	Triplex receptacle outlet

Unless noted, assume every receptacle will be grounded, and will have a separate grounding contact.

RECEPTACLE OUTLET SYMBOLS (*cont.*)

 Quadruplex receptacle outlet

 Duplex receptacle outlet-split wired

 Triplex receptacle outlet-split wired

 Single special-purpose receptacle outlet*

 Duplex special-purpose receptacle outlet*

 Range outlet

* Use numeral or letter, either within the symbol or as a subscript alongside the symbol keyed to explanation in the drawing list of symbols, to indicate type of receptacle or usage.

 Special-purpose connection or provision for connection. Use subscript letters to indicate function (DW—dishwasher; CD—clothes dryer, etc.)

 Multioutlet assembly. Extend arrows to limit of installation. Use appropriate symbol to indicate type of outlet. Also indicate spacing of outlets as x inches.

RECEPTACLE OUTLET SYMBOLS (*cont.*)

Symbol	Description
(C)	Clock hanger receptacle
(F)	Fan hanger receptacle
⊟	Floor single receptacle outlet
⊟	Floor duplex receptacle outlet
◬	Floor special-purpose outlet*
◀	Floor telephone outlet – public
◁	Floor telephone outlet – private
▷ ▶ ⊖	Not a part of the Standard: example of the use of several floor outlet symbols to identify a 2-, 3-, or more-gang floor outlet
⊞	Underfloor duct and junction box for triple, double or single duct system indicated by number of parallel lines

RECEPTACLE OUTLET SYMBOLS (*cont.*)

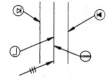

Not a part of the Standard: example of use of various symbols to identify location of different types of outlets or connections for underfloor duct or cellular floor systems

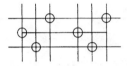

Cellular floor header duct

* Use numeral or letter, either within the symbol or as a subscript alongside the symbol keyed to explanation in the drawing list of symbols, to indicate type of receptacle or usage.

SWITCH OUTLET SYMBOLS

S	Single-pole switch
S_2	Double-pole switch
S_3	Three-way switch
S_4	Four-way switch
S_K	Key-operated switch
S_P	Switch and pilot lamp

SWITCH OUTLET SYMBOLS

S_L — Switch for low-voltage switching system

S_{LM} — Master switch for low-voltage switching system

⊖ S — Switch and single receptacle

⊖ S — Switch and double receptacle

S_D — Door switch

S_T — Time switch

S_{CB} — Circuit-breaker switch

S_{MC} — Momentary contact switch or push-button for other than signaling system

8-8

SIGNALING SYSTEM OUTLET SYMBOLS FOR INSTITUTIONAL, COMMERCIAL AND INDUSTRIAL OCCUPANCIES

These symbols are recommended by the American Standards Association, but are not used universally. The reader should remember not to assume that these symbols will be used on any certain plan, but to always check the symbol list on the plans, and verify if these symbols are actually used.

Basic Symbol	Examples of Individual Item Identification with description (Not part of the standard)
⊸◯	*Nurse Call System Devices (any type)*
⊸① 12	Nurses' Annunciator (can add a number after it as to indicate number of lamps) EX:12
⊸②	Call station, single cord, pilot light
⊸③	Call station, double cord, microphone speaker
⊸④	Corridor dome light, 1 lamp
⊸⑤	Transformer
⊸⑥	Any other item on same system- use numbers as required.

Basic Symbol	Examples of Individual Item Identification with description (Not part of the standard)

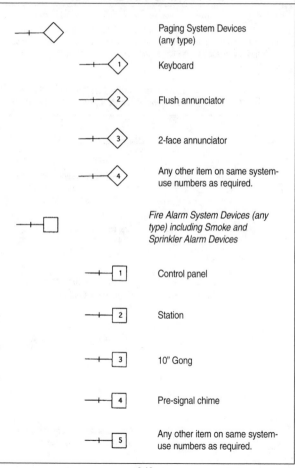

Paging System Devices (any type)

Keyboard

Flush annunciator

2-face annunciator

Any other item on same system- use numbers as required.

Fire Alarm System Devices (any type) including Smoke and Sprinkler Alarm Devices

Control panel

Station

10" Gong

Pre-signal chime

Any other item on same system- use numbers as required.

Basic Symbol	Examples of Individual Item Identification with description (Not part of the standard)

Staff Register System Devices
(any type)

Phone operators' register — (1)

Entrance register–flush — (2)

Staff room register — (3)

Transformer — (4)

Any other item on same system-use numbers as required. — (5)

Electric Clock System Devices
(any type)

Master clock — (1)

12" Secondary–flush — (2)

12" Double dial–wall mounted — (3)

18" Skeleton dial — (4)

Any other item on same system-use numbers as required. — (5)

Basic Symbol	Examples of Individual Item Identification with description (Not part of the standard)
──┼─◁	*Public Telephone System Devices*
──┼─◁ 1	Switchboard
──┼─◁ 2	Desk phone
──┼─◁ 3	Any other item on same system-use numbers as required.
──┼┼─◀	*Private Telephone System Devices (any type)*
──┼┼─◀ 1	Switchboard
──┼┼─◀ 2	Wall phone
──┼┼─◀ 3	Any other item on same system-use numbers as required.
──┼─⌂	*Watchman System Devices (any type)*
──┼─⌂ 1	Central station
──┼─⌂ 2	Key station
──┼─⌂ 3	Any other item on same system-use numbers as required.

Basic Symbol	Examples of Individual Item Identification with description (Not part of the standard)
⊢⊣◁	*Sound System*
⊢⊣◁1	Amplifier
⊢⊣◁2	Microphone
⊢⊣◁3	Interior speaker
⊢⊣◁4	Exterior speaker
⊢⊣◁5	Any other item on same system-use numbers as required.
⊢⊣◯	*Other Signal System Devices*
⊢⊣①1	Buzzer
⊢⊣②2	Bell
⊢⊣③3	Pushbutton
⊢⊣④4	Annunciator
⊢⊣⑤5	Any other item on same system-use numbers as required.

SIGNALING SYSTEM OUTLET SYMBOLS
FOR RESIDENTIAL OCCUPANCIES

When a descriptive symbol list is not employed, use the following signaling system symbols to identify standardized, residential-type, signal-system items on residential drawings.

Symbol	Description
▣	Pushbutton
◁▢	Buzzer
◖▢	Bell
◖▢╱	Combination bell-buzzer
CH	Chime
◇─	Annunciator
D	Electric door opener
M	Maid's signal plug
▢	Interconnection box
BT	Bell-ringing transformer
▶	Outside telephone
▷	Interconnecting telephone
R	Radio outlet
TV	Television outlet

PANELBOARDS, SWITCHBOARDS AND RELATED EQUIPMENT SYMBOLS

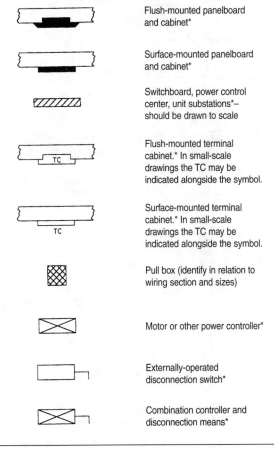

Flush-mounted panelboard and cabinet*

Surface-mounted panelboard and cabinet*

Switchboard, power control center, unit substations*– should be drawn to scale

Flush-mounted terminal cabinet.* In small-scale drawings the TC may be indicated alongside the symbol.

Surface-mounted terminal cabinet.* In small-scale drawings the TC may be indicated alongside the symbol.

Pull box (identify in relation to wiring section and sizes)

Motor or other power controller*

Externally-operated disconnection switch*

Combination controller and disconnection means*

*Identify by notation or schedule.

BUS DUCTS AND WIREWAYS SYMBOLS

| T | T | T | Trolley duct* |

Busway (service, feeder, or plug-in)*

| B | B | B |

| C | C | C | Cable trough ladder or channel* |

| W | W | W | Wireway* |

REMOTE CONTROL STATION SYMBOLS FOR MOTORS OR OTHER EQUIPMENT

Pushbutton station

F Float switch–mechanical

L Limit switch–mechanical

P Pneumatic switch–mechanical

Electric eye–beam source

Electric eye–relay

T Thermostat

Identify by notation or schedule.

CIRCUITING SYMBOLS

Wiring method identification by notation on drawing or in specification.

——————————————— Wiring concealed in ceiling or wall

- - - - - - - - - - - - Wiring concealed in floor

· · · · · · · · · · · · · · Wiring exposed

Note: Use heavyweight line to identify service and feeders. Indicate empty conduit by notation CO (conduit only).

2   1
————————→ Branch-circuit home run to panelboard. Number of arrows indicates number of circuits. (A numeral at each arrow may be used to identify circuit number.) Note: Any circuit without further identification indicates two-wire circuit. For a greater number of wires, indicate with cross lines.

3 wires
————/ / /————

4 wires
————/ / / /———— Unless indicated otherwise, the wire size of the circuit is the minimum size required by the specification.

Identify different functions of wiring system, e.g., signaling system by notation or other means.

————————○ Wiring turned up

————————● Wiring turned down

ELECTRIC DISTRIBUTION OR
LIGHTING SYSTEM – UNDERGROUND

Symbol	Description
M	Manhole*
H	Handhole*
TM	Transformer manhole or vault*
TP	Transformer pad*
– – – – –	Underground direct burial cable. Indicate type, size, and number of conductors by notation or schedule.
	Underground duct line. Indicate type, size, and number of ducts by cross-section identification of each run by notation or schedule. Indicate type, size, and number of conductors by notation or schedule.
	Streetlight standard feed from underground circuit*

* Identify by notation or schedule.

8-18

ELECTRIC DISTRIBUTION OR
LIGHTING SYSTEM – AERIAL

Pole*

Streetlight and bracket*

Transformer*

Primary circuit*

Secondary circuit*

Down guy

Head guy

Sidewalk guy

Service weather head*

Identify by notation or schedule.

8-19

ARRESTER, LIGHTING ARRESTER GAP SYMBOLS (ELECTRIC SURGE)

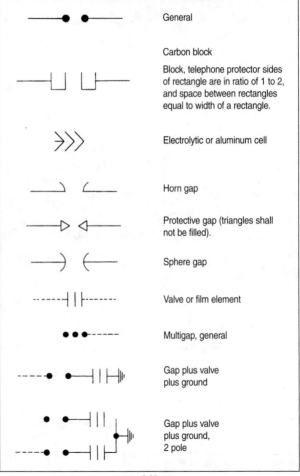

General

Carbon block

Block, telephone protector sides of rectangle are in ratio of 1 to 2, and space between rectangles equal to width of a rectangle.

Electrolytic or aluminum cell

Horn gap

Protective gap (triangles shall not be filled).

Sphere gap

Valve or film element

Multigap, general

Gap plus valve plus ground

Gap plus valve plus ground, 2 pole

BATTERY SYMBOLS

The long line is always positive, but polarity may also be indicated as shown in the direct-current source.

  Generalized direct-current source

One cell

Multicell

Multicell battery with 3 taps

Multicell battery with adjustable tap

CIRCUIT BREAKER SYMBOLS

If it is desired to show the condition causing the breaker to trip, the relayprotective-function symbols may be used alongside the break symbol. On a power diagram, the symbol may be used without other indentification. On a composite drawing the identifying letter CB inside or adjacent to the square may be added.

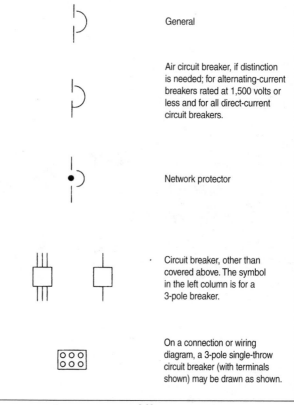

General

Air circuit breaker, if distinction is needed; for alternating-current breakers rated at 1,500 volts or less and for all direct-current circuit breakers.

Network protector

Circuit breaker, other than covered above. The symbol in the left column is for a 3-pole breaker.

On a connection or wiring diagram, a 3-pole single-throw circuit breaker (with terminals shown) may be drawn as shown.

CIRCUIT BREAKER APPLICATION SYMBOLS

3-pole circuit breaker

3-pole circuit breaker with thermal overload device in all 3 poles

3-pole circuit breaker with magnetic overload device in all 3 poles

3-pole circuit breaker, drawout type

CIRCUIT RETURN SYMBOLS

Ground

a) A direct conducting connection to the earth or body of water

b) A conducting connection to a structure similar to an earth ground (frame of an air, space, or vehicle that is not conductively connected to earth)

Chassis or frame connection

A conducting connection to a chassis or frame of a unit. The chassis or frame may be a substantial potential with respect to the earth or structure in which this chassis or frame is mounted.

Common connections

Conducting connections made to one another. All like-designated points are connected. *The asterisk is not part of the symbol. Identifying valves, letters, numbers, or marks shall replace the asterisk.

CONTACTOR SYMBOLS

Symbol	Description
	Fixed contact for jack, key, relay, etc.
	Fixed contact for switch
	Fixed contact for momentary switch
	Sleeve
	Adjustable or sliding contact for resistor, inductor, etc.
	Locking
	Segment; bridging contact
	Nonlocking
	Vibrator reed
	Vibrator split reed
	Rotating contact (slip ring) and brush

8-25

CIRCUIT CONDITION DESIGNATIONS

It is standard procedure to show a contact by a symbol which indicates the circuit condition produced when the actuating device is in the nonoperated, or de-energized, position. On many wiring diagrams it is necessary to add a clarifying note explaining the proper point at which the contact functions. That is, the point where the electrical or mechanical device opens or closes due to operating changes in pressure, level, flow, voltage, current, etc. When it is necessary to show contacts in the operated or energized condition, a clarifying note shall be added to the diagram. For example, contacts for circuit breakers or auxiliary switches may be designated as shown below:

    (a) Closed when device is in energized or operated position

    (b) Closed when device is in de-energized or nonoperated position

    (aa) Closed when operating mechanism or main device is in energized or operated position

    (bb) Closed when operating mechanism of main device is in de-energized or nonoperated position

PARALLEL-LINE CONTACTOR SYMBOLS

In the parallel-line contact, symbols showing the length of the parallel lines shall be approximately 1-1/4 times the width of the gap (except for Time Sequential Closing).

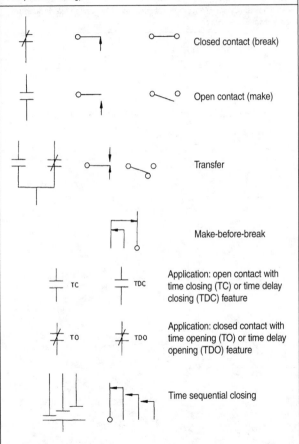

Closed contact (break)

Open contact (make)

Transfer

Make-before-break

Application: open contact with time closing (TC) or time delay closing (TDC) feature

Application: closed contact with time opening (TO) or time delay opening (TDO) feature

Time sequential closing

OPERATED CONTACTOR SYMBOLS

Contactor symbols are derived from fundamental contact, coil, and mechanical connection symbols. A complete diagram of the actual contactor device is constructed by combining fundamental symbols for mechanical connections, control circuits, etc. Mechanical interlocking will be indicated by notes.

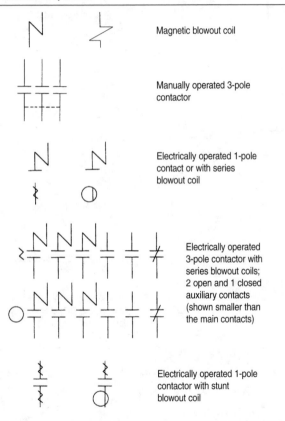

Magnetic blowout coil

Manually operated 3-pole contactor

Electrically operated 1-pole contact or with series blowout coil

Electrically operated 3-pole contactor with series blowout coils; 2 open and 1 closed auxiliary contacts (shown smaller than the main contacts)

Electrically operated 1-pole contactor with stunt blowout coil

GENERATOR AND MOTOR SYMBOLS

◯	Basic
(GEN)	Generator
(MOT)	Motor
	Rotating armature with commutator and brushes
⌒	Compensating or commutating
⌒⌒	Series
⌒⌒⌒	Shunt, or separately excited magnet, permanent

WINDING SYMBOLS

Generator and motor winding symbols may be shown in the basic circle using the following representation.

⊘	1-phase
⊗	2-phase

WINDING SYMBOLS (cont.)

 3-phase wye (ungrounded)

 3-phase wye (grounded)

 3-phase delta

 6-phase diametrical

 6-phase double delta

APPLICATIONS FOR DIRECT-CURRENT

  Separately excited direct-current generator or motor

  Separately excited direct-current generator or motor; with commutating or compensating field winding or both

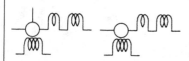 Compositely excited direct-current generator or motor; with commutating or compensating field winding or both

8-30

APPLICATIONS FOR DIRECT-CURRENT (*cont.*)

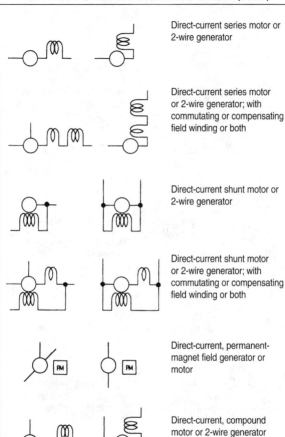

Direct-current series motor or 2-wire generator

Direct-current series motor or 2-wire generator; with commutating or compensating field winding or both

Direct-current shunt motor or 2-wire generator

Direct-current shunt motor or 2-wire generator; with commutating or compensating field winding or both

Direct-current, permanent-magnet field generator or motor

Direct-current, compound motor or 2-wire generator or stabilized shunt motor

APPLICATIONS FOR DIRECT-CURRENT (*cont.*)

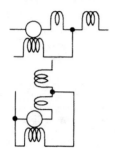

Direct-current compound motor or 2-wire generator or stabilized shunt motor; with commutating or compensating field winding or both

Direct-current, 3-wire shunt generator

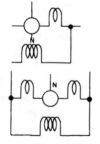

Direct-current, 3-wire shunt generator; with commutating or compensating field winding or both

APPLICATIONS FOR DIRECT-CURRENT (*cont.*)

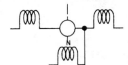

Direct- current, 3-wire compound generator*

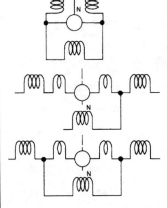

Direct-current, 3-wire compound generator; with commutating or compensating field winding or both*

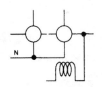

Direct-current balancer, shunt wound*

---

* The broken line — · — indicates where line connection to a symbol is made and is not a part of the symbol.

APPLICATIONS FOR DIRECT-CURRENT (cont.)

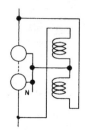

Direct-current balancer, shunt wound.*

ALTERNATING CURRENT MOTOR SYMBOLS

Squirrel-cage induction motor or generator, split-phase induction motor or generator, rotary phase converter or repulsion motor

Wound-rotor induction motor, synchronous induction motor, induction generator, or induction frequency converter

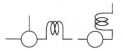

Alternating-current series motor*

The broken line — · — indicates where line connection to a symbol is made and is not a part of the symbol.

PATH – TRANSMISSION, CONDUCTOR, CABLE, WIRING, ETC.

The entire group of conductors, or the transmission path required to guide the power or symbol, is shown by a single line. In coaxial and waveguide work, the recognition symbol is employed at the beginning and end of each type of transmission path as well as at intermediate points to clarify a potentially confusing diagram. For waveguide work, the mode may be indicated as well.

General; conductive, conductor or wire guided conductor or wire path

Two conductors or conductive paths of wires

Three conductors or conductive paths of wires

"n" conductors or conductive paths of wires

Crossing of paths or conductors not connected

The crossing is not necessarily at a 90-degree angle.

JUNCTION OF PATHS/CONDUCTORS SYMBOLS

Junction of paths or conductors

Application: junction of paths, conductor, or cable. Path type or size may be indicated on diagram

Splice

Application: splice of same size cables.

Junction of conductors of same size or different size cables. Sizes of conductors should be indicated on diagram

Junction of connected paths, conductors, or wires

POLARITY SYMBOLS

\+   Positive

–   Negative

SWITCH SYMBOLS

Switch symbols are constructed of basic symbols for mechanical connections, contacts, etc. and normally a switch is represented in the de-energized position for switches having two or more positions where no operating force is applied. When actuated by a mechanical force, the functioning point is described by a clarifying note. Where switch symbols are in closed position, the terminals should be included for clarity.

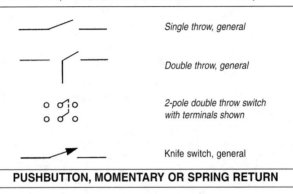

Single throw, general

Double throw, general

2-pole double throw switch with terminals shown

Knife switch, general

PUSHBUTTON, MOMENTARY OR SPRING RETURN

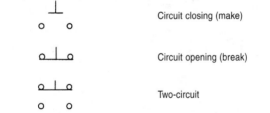

Circuit closing (make)

Circuit opening (break)

Two-circuit

PUSHBUTTON, MAINTAINED OR NOT SPRING RETURN

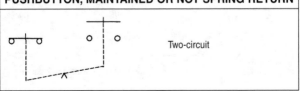

Two-circuit

TRANSFORMER SYMBOLS

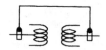

Transformer with direct-current connections and mode suppression between two rectangular waveguides

Transformer with Magnetic-core shown

Shielded transformer with magnetic core shown

Transformer with magnetic core shown and with a shield between windings. The shield is shown connected to the frame

With taps, 1-phase

Autotransformer, 1-phase

Adjustable

1-phase, 2-winding transformer

TRANSFORMER SYMBOLS (*cont.*)

  3-phase bank of 1-phase, 2-winding transformer

  Polyphase transformer

 Current transformer(s)

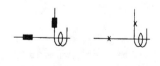 Current transformer with polarity marking. Instantaneous direction of current into one polarity mark responds to current out of the other polarity mark.

TRANSFORMER SYMBOLS (cont.)

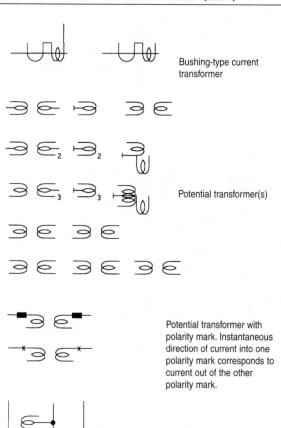

Bushing-type current transformer

Potential transformer(s)

Potential transformer with polarity mark. Instantaneous direction of current into one polarity mark corresponds to current out of the other polarity mark.

Outdoor metering device

TRANSFORMER WINDING CONNECTION SYMBOLS

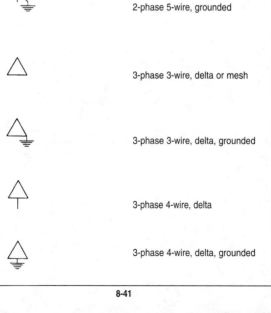

2-phase 3-wire

2-phase 3-wire, grounded

2-phase 4-wire

2-phase 5-wire, grounded

3-phase 3-wire, delta or mesh

3-phase 3-wire, delta, grounded

3-phase 4-wire, delta

3-phase 4-wire, delta, grounded

TRANSFORMER WINDING CONNECTION SYMBOLS (*cont.*)

3-phase, open-delta

3-phase, open-delta, grounded at common point

3-phase, open-delta, grounded at middle point of one transformer

3-phase, broken-delta

3-phase, wye or star

3-phase, wye, grounded neutral

3-phase 4-wire

MOTOR CONTROL SYMBOLS

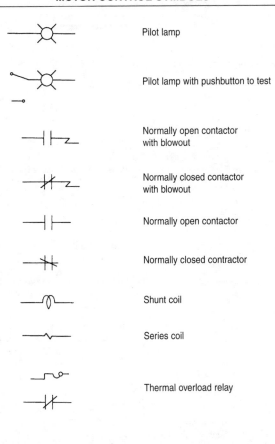

Pilot lamp

Pilot lamp with pushbutton to test

Normally open contactor
with blowout

Normally closed contactor
with blowout

Normally open contactor

Normally closed contractor

Shunt coil

Series coil

Thermal overload relay

Magnetic relay

MOTOR CONTROL SYMBOLS (*cont.*)

Limit switch, normally open

Limit switch, normally closed

Footswitch, normally open

Footswitch, normally closed

Vacuum switch, normally open

Vacuum switch, normally closed

Liquid-level switch, normally open

Liquid-level switch, normally closed

Temperature-actuated switch, normally open

Temperature-actuated switch, normally closed

Flow switch, normally open

Flow switch, normally closed

MOTOR CONTROL SYMBOLS (*cont.*)

Symbol	Description
	Momentary-contact switch, normally open
	Momentary-contact switch, normally closed
	Iron-core inductor
	Air-core inductor
	Single-phase AC motor
	3-phase, squirrel-cage motor
	2-phase, 4-wire motor
	Wound-rotor, 3-phase motor
	Armature
	Crossed wires, not connected
	Crossed wires, connected
	Fuse

MOTOR CONTROL SYMBOLS (*cont.*)

	Thermocouple
	Diode (rectifier)
	Capacitor
	Adjustable capacitor
	Resistor
	Tapped resistor
	Variable resistor
	Resistor with two taps
	Wiring terminal
	Full-wave rectifier
	Mechanical interlock
	Mechanical connection

METER AND INSTRUMENT SYMBOLS

A	Ammeter IEC
AH	Ampere-hour
CMA	Contact-making (or breaking) ammeter
CMC	Contact-making (or breaking) clock
CMV	Contact-making (or breaking) voltmeter
CRO	Oscilloscope or cathode-ray oscillograph
DB	DB (decibel) meter
DBM	DMB (decibels referred to 1 milliwatt) meter
DM	Demand meter
DTR	Demand-totalizing relay
F	Frequency meter
G	Glavanometer
GD	Ground detector
I	Indicating
INT	Integrating
μA or UA	Microammeter
MA	Milliammeter
NM	Noise meter
OHM	Ohmmeter
OP	Oil pressure

METER AND INSTRUMENT SYMBOLS (cont.)

OSCG	Oscillograph string
PH	Phasemeter
PI	Position indicator
PF	Power factor
RD	Recording demand meter
REC	Recording
RF	Reaction factor
SY	Synchroscope
TLM	Telemeter
T	Temperature meter
THC	Thermal converter
TT	Total time
V	Voltmeter
VA	Volt-ammeter
VAR	Varmeter
VARH	Varhour meter
VI	Volume indicator; meter, audio level
VU	Standard volume indicator; meter, audio level
W	Wattmeter
WH	Watthour meter

CHAPTER 9
LIGHTING

RECOMMENDED LIGHT LEVELS

Area	Footcandles
Hospital operating table	2500
Factory assembly area	100-200
Accounting office	150
Major league baseball infield	150
Major league baseball outfield	100
School classroom	60-90
General office	50-80
Home kitchen	50-70
Bank lobby	50
Conference rooms	30-50
Active warehouse storage area	20-30
Corridors	10-20
Auditorium	15
Mechanical space	15
Elevator	10
Parking lot	5
Inactive warehouse storage area	5
Interstate roadway	1.4
Street	.9

LAMP RATINGS (W=WATTS)

Lamp	Initial Lumen	Mean Lumen
40 W standard incandescent	480	N/A
100 W standard incandescent	1750	N/A
40 W standard fluorescent	3400	3100
100 W tungsten-halogen	1800	1675
100 W mercury-vapor	4000	3000
250 W mercury-vapor	12,000	9800
100 W high-pressure sodium	9500	8500
250 W high-pressure sodium	30,000	27,000
250 W metal-halide	20,000	17,000

LAMP ADVANTAGES AND DISADVANTAGES

Lamp	Advantages	Disadvantages
Incandescent, tungsten-halogen	Low initial cost Simple construction No ballast required Available in many shapes and sizes Requires no warm-up or restart time Inexpensively dimmed Simple maintenance	Low electrical efficiency High operating temperature Short life Bright light source in small space Does not allow large distribution of light
Fluorescent	Available in many shapes and sizes Moderate cost Good electrical efficiency Long life Low shadowing Low operating temperature Short turn-ON delay	Not suited for high-level light in small, highly-concentrated applications Requires ballast Higher initial cost than incandescent lamps Light output and color affected by ambient temperature Expensive to dim
Low-pressure sodium, mercury-vapor, metal-halide, high-pressure	Good electrical efficiency Long life High light output Slightly affected by ambient temperature	May cause color distortion Long start and restart time High initial cost High replacement cost Requires ballast Expensive or not possible to dim Problem starting in cold weather High-socket voltage required

LAMP ELECTRICAL EFFICIENCY

Lamp	Lumen Per Watt*
Incandescent	15-25
Mercury-vapor	50-60
Fluorescent	55-100
Metal-halide	80-125
High-pressure sodium	80-150
Low-pressure sodium	160-200

* exact lumen output depends on size and type of lamp used

COMMON INCANDESCENT LAMPS

Type	Watts	Size (inches)	Volts
A-19	40-100	2-3/8	120-130
F-15	25-60	1-7/8	120
PS-35	300-500	4-3/8	120
PAR-38	115-150	4-3/4	120-130
T-19	40-100	2-3/8	120

COMMON FLUORESCENT LAMPS

Type	Watts	Length (inches)	Size (inches)	Base
T-5	4-13	6-21	5/8	Miniature Bi-pin
T-5	13	21	5/8	Miniature Bi-pin
T-8	15-30	18-36	1	Medium Bi-pin
T-9	20-40	6-16	1-1/8	4-pin
T-10	40	48	1-1/4	Medium Bi-pin
T-12	14-75	15-96	1-1/2	Medium Bi-pin
T-12	60-75	96	1-1/2	Single Bi-pin

BALLAST SOUND RATING

Sound Rating	Noise Level (decibels)	Recommended Application
A	20-24	Broadcasting booths, churches, study areas
B	25-30	Libraries, classrooms, quiet office areas
C	31-36	General office areas, commercial buildings
D	37-42	Retail stores, stock rooms
E	43-48	Light production areas, general outdoor lighting
F	Over 48	Street lighting, heavy production areas

HID LAMP OPERATING CHARACTERISTICS

Lamp	Start Time (minutes)	Restart Time
Low-pressure sodium	6-12	4-12 sec.
Mercury-vapor	5-6	3-5 min.
Metal-halide	2-5	10-15 min.
High-pressure sodium	3-4	30-60 sec.

LAMP CHARACTERISTICS SUMMARY

Lamp	Lm/W	Bulb Life (hours)	Color Rendition	Operating Cost
Incandescent	15-25	750-1000	Excellent	Very high
Tungsten-halogen	20-25	1500-2000	Excellent	High
Fluorescent	55-100	7500-24,000	Very good	Average
Low-press. sodium	190-200	1800	Poor	Low
Mercury-vapor	50-60	16,000-24,000	Varys on type used	Average
Metal-halide	80-125	3000-20,000	Very good	Average
High-press. sodium	65-115	7500-14,000	Good (golden white)	Low

HID LAMP TROUBLESHOOTING GUIDE

Problem	Possible Cause	Corrective Action
Lamp does not start	Normal end of life operating characteristic	Replace with new lamp.
	Loose lamp connection	Re-seat lamp in socket securely. Ensure lamp holder is rigidly mounted and properly spaced.
	Defective photocell used for automatic turn ON	Replace defective photocell.
	Low line voltage applied to circuit	Ensure that voltage is within ±7% of rated voltage.
	Cold area or draft hitting lamp	Protect lamp with enclosure. Use a low temperature-rated ballast.
	Defective ballast	Replace ballast.
Lamp cycles ON and OFF or flickers when ON	Normal end of life operating characteristic	Replace with new lamp.
	Poor electrical connection or loose bulb	Check electrical connections and socket contacts.
	Line voltage variations	Ensure that voltage is within ±7% of rated voltage. Move lamps to separate circuit when lamps are on same circuit as high-power loads. High overloads cause a voltage dip when turned ON. This voltage dip may cause the lamp to turn OFF.
Short lamp life for line voltage	Wrong wattage lamp or ballast	HID lamps and ballast must be matched in size. Lamp life is shortened when a large-wattage lamp is used. The same size and type ballast must be installed when replacing a ballast.
	Power tap set too low for the correct supply voltage	Check ballast taps and ensure they are set for the correct supply voltage.
	Defective sodium ballast starter	Replace with new lamp.
Low light output	Normal end of life operating characteristic	Replace with new lamp.
	Dirty lamp or lamp fixture	Keep lamp fixture clean.
	Early blackening of bulb caused by incorrect lamp size or ballast	Ensure that lamp size, lamp type, and ballast match

INCANDESCENT LAMP TROUBLESHOOTING GUIDE

Problem	Possible Cause	Corrective Action
Short lamp life	Voltage higher than lamp rating	Ensure that voltage is equal to or less than lamp's rated voltage. Lamp life is decreased when the voltage is higher than the rated voltage. A 5% higher voltage shortens the lamp life by 40%.
	Lamp exposed to rough service conditions or vibration	Replace lamp with one rated for rough service or resistant to vibration.
Lamp does not turn ON after new lamp is installed	Fuse blown, poor electrical connection, faulty control switch	Check circuit fuse or CB. Check electrical connections and voltage out of control switch. Replace switch when voltage is present into, but not out of the control switch.

FLUORESCENT LAMP TROUBLESHOOTING GUIDE

Problem	Circuit Type	Possible Cause	Corrective Action
Lamp blinks & has shimmering effect during lighting period	All types	Depletion of emission material on electrodes	Replace with new lamp.
New lamp blinks	All types	Loose lamp connection	Reseat lamp in socket securely. Ensure lamp holders are rigidly mounted and properly spaced.
		Low voltage applied to circuit	Ensure that voltage is within ±7% of rated voltage.
		Cold area or draft hitting lamp	Protect lamp with enclosure. Use a low temperature-rated ballast.
	Preheat	Defective or old starter	Replace starter. Lamp life is reduced when starter is not replaced.
Lamp does not light or is slow starting	All types	Lamp failure	Replace lamp. Test faulty lamp in another fixture. Check circuit fuse or CB. Check voltage at fixture.
		Loss of power or low voltage to the fixture	Troubleshoot fluorescent fixture when voltage is present at the fixture. Replace broken or cracked holder. Check for poor wire connection.
	Preheat Rapid-start	Normal end of starter life Failed capacitor in ballast	Replace starter. Replace ballast.

FLUORESCENT LAMP TROUBLESHOOTING GUIDE (cont.)

Problem	Circuit Type	Possible Cause	Corrective Action
Lamp does not light	All types	Lamp not seated in holder	Seat lamp properly in holder. In rapid-start circuits, the holder includes a switch that removes power when the lamp is removed due to high voltage present.
Bulb ends remain lighted after switch is turned OFF	Preheat	Starter contacts stuck together	Replace starter. Lamp life is reduced when starter is not replaced.
Short lamp life	All types	Frequent turning ON and OFF of lamps	Normal operation is based on one start per three hour period of operation time. Short lamp life must be expected when frequent starting cannot be avoided.
		Supply voltage excessive or low	Check the supply voltage against the ballast rating. Short lamp life must be expected when supply voltage is not within ±7% of lamp rating.
		Low ambient temperature. Low temperature causes a slow start	Protect lamp with enclosure. Use a low temperature-rated ballast.
	Instant-start	One lamp burned out and other burning dimly due to series-start ballast circuit	Replace burned-out lamp. Ballast is damaged when lamp is not replaced.
		Wrong lamp type. May be using rapid-start or pre-heat lamp instead of instant-start	Replace lamp with correct type.
Light output decreases	All types	Light output decreases over first 100 hours of operation	Rated light output is based on output after 100 hours of lamp operation. Before 100 hours of operation, light output may be as much as 10% higher than normal.
		Low circuit voltage	Check the supply voltage against the ballast rating. Short lamp life and low light output must be expected when supply voltage is not within ±7% of lamp rating.
		Dirt build-up on lamp and fixture	Clean bulb and fixture.

LAMP IDENTIFICATION			
Lamp	Letter	Number	Rating
Mercury-vapor	H	33	400 W
		34	1000 W/130 V
		35	700 W
		36	1000 W/265 V
		37	250 W
		38	100 W
		39	175 W
		43	75 W
Self-ballasted Mercury-vapor	B	78	750 W/120 V
Metal-halide	M	47	1000W
		48	1500 W
		57	175 W
		58	250 W
		59	400 W
High-pressure sodium	S	50	250 W
		51	400 W
		52	1000 W
		54	100 W
		55	150 W/55 V
		56	150 W/100 V
		62	70 W
		66	200 W
		67	310 W
		68	50 W
		76	35 W

BALLAST LOSS		
Lamp	Lamp Rated Wattage (in W)	Ballast Loss (power loss %)
Low-pressure sodium	70	27
	100	25
	150	22
	250	20
	400	15
	1000	7
Mercury-vapor	40	32
	75	25
	100	22
	175	17
	250	16
	400	14
	700	7
	1000	7
Metal-halide	175	20
	250	17
	400	13
	1000	7
	1500	7
High-pressure sodium	50	30
	100	24
	150	22
	250	20
	400	16
	1000	7

RECOMMENDED BALLAST OUTPUT VOLTAGE LIMITS

Ballast	Lamp Size		RMS
	Wattage	ANSI Number	Voltage (Volts)
Low-pressure sodium	18	L69	300-325
	35	L70	455-505
	55	L71	455-505
	90	L72	455-525
	135	L73	645-715
	180	L74	645-715
Mercury-vapor	50	H46	225-255
	75	H43	225-255
	100	H38	225-255
	175	H39	225-255
	250	H37	225-255
	400	H33	225-255
	700	H35	405-455
	1000	H36	405-455
Metal-halide	70	M85	210-250
	100	M90	250-300
	150	M81	220-260
	175	M57	285-320
	250	M80	230-270
	250	M58	285-320
	400	M59	285-320
	1000	M47	400-445
	1500	M48	400-445
High-pressure sodium	35	S76	110-130
	50	S68	110-130
	70	S62	110-130
	100	S54	110-130
	150	S55	110-130
	150	S56	200-250
	200	S66	200-230
	250	S50	175-225
	310	S67	155-190
	400	S51	175-225
	1000	S52	420-480

RECOMMENDED SHORT-CIRCUIT CURRENT TEST LIMITS

Ballast	Lamp Size		Short-Circuit Current Range (Amps)
	Wattage	ANSI Number	
Low-pressure sodium	18	L69	.30 - .40
	35	L70	.52 - .78
	55	L71	.52 - .78
	90	L72	.8 - 1.2
	135	L73	.8 - 1.2
	180	L74	.8 - 1.2
Mercury-vapor	50	H46	.85 - 1.15
	75	H43	.95 - 1.70
	100	H38	1.10 - 2.00
	175	H39	2.0 - 3.6
	250	H37	3.0 - 3.8
	400	H33	4.4 - 7.9
	700	H35	3.9 - 5.85
	1000	H36	5.7 - 9.0
Metal-halide	70	M85	.85 - 1.30
	100	M90	1.15 - 1.76
	150	M81	1.75 - 2.60
	175	M57	1.5 - 1.90
	250	M80	2.9 - 4.3
	250	M58	2.2 - 2.85
	400	M59	3.5 - 4.5
	1000	M47	4.8 - 6.15
	1500	M48	7.4 - 9.6
High-pressure sodium	35	S76	.85 - 1.45
	50	S68	1.5 - 2.3
	70	S62	1.6 - 2.9
	100	S54	2.45 - 3.8
	150	S55	3.5 - 5.4
	150	S56	2.0 - 3.0
	200	S66	2.50 - 3.7
	250	S50	3.0 - 5.3
	310	S67	3.8 - 5.7
	400	S51	5.0 - 7.6
	1000	S52	5.5 - 8.1

LIGHT SOURCE CHARACTERISTICS

Characteristics	Incandescent, Including Tungsten	Fluorescent	High-Intensity Discharge			
			Mercury Vapor (Self-Ballasted)	Metal Halide	High-Pressure Sodium (Improved Color)	Low-Pressure Sodium
Wattages (lamp only)	15-1500	15-219	40-1000	175-1000	70-1000	35-180
Life* (hr.)	750-12,000	7500-24,000	16,000-15,000	1500-15,000	24,000 (10,000)	18,000
Efficacy* (lumens/W) lamp only	15-25	55-100	50-60 (20-25)	80-100	75-140 (67-112)	Up to 180
Lumen maintenance	Fair to excellent	Fair to excellent	Very good (good)	Good	Excellent	Excellent
Color rendition	Excellent	Good to excellent	Poor to excellent	Very good	Fair	Good
Light direction control	Very good to excellent	Fair	Very good	Very good	Very good	Fair
Source size	Compact	Extended	Compact	Compact	Compact	Extended
Relight time	Immediate	Immediate	3-10 min.	10-20 min.	Less than 1 min.	Immediate
Comparative fixture cost	Low: simple fixtures	Moderate	Higher than incandescent and fluorescent	Generally higher than mercury	High	High
Comparative operating cost	High: short life and low efficiency	Lower than incandescent	Lower than incandescent	Lower than mercury	Lowest of HID types	Low
Auxiliary equipment needed	Not needed	Needed: medium cost	Needed: high cost	Needed: high cost	Needed: high cost	Needed: high cost

*Life and efficacy ratings subject to revision. Check manufacturers' data for latest information.

FUNDAMENTAL LIGHTING CALCULATIONS

AVERAGE ILLUMINANCE CALCULATION

There are two basic types of lighting calculations. The first is the AVERAGE ILLUMINANCE CALCULATION, which predicts the average footcandles in a space.

Footcandles are lumens of light per square foot of area. Assume that 1000 lumens of light are evenly distributed over an area of 100 square feet. The illuminance will be calculated as follows:

1000 LM/100 SQ. FT., OR 10 LUMENS PER SQUARE FOOT, WHICH EQUALS 10 FOOTCANDLES.

Note: The percentage of total lamp lumens which actually reach a work surface will be less than the total number of installed lumens, typically ranging from 40% to 70%.

INVERSE SQUARE LAW

The second calculation method is called the INVERSE SQUARE LAW, which predicts illuminance at a specific point in a space. It is the most commonly used and forms the basis for photometric testing of luminaries. As light travels away from a source it spreads out and is distributed over a wider area reducing its density.

The reduction is equal to $1/distance^2$, thus the name "inverse square law." The illuminance (foot candles) at a point which is at right angles to the fixture may be found from the formula below:

$$\text{FOOT CANDLES AT A PARTICULAR POINT} = \frac{\text{CANDLEPOWER (CANDELAS)}}{\text{DISTANCE}^2}$$

For example, if the intensity is 1000 candelas and the distance is 10 feet, the illuminance will be 1000 candelas/100 sq. ft. or 10 foot candles.

Note: If the surface is not at a right angle the equation must be modified to reflect the further spread of light due to the angle.

CHAPTER 10
CONVERSION FACTORS AND UNITS OF MEASUREMENT

COMMONLY USED CONVERSION FACTORS

Multiply	By	To Obtain
Acres	43,560	Square feet
Acres	1.562×10^{-3}	Square miles
Acre-Feet	43,560	Cubic feet
Amperes per sq cm	6.452	Amperes per sq in.
Amperes per sq in.	0.1550	Amperes per sq cm
Ampere-Turns	1.257	Gilberts
Ampere-Turns per cm	2.540	Ampere-turns per in.
Ampere-Turns per in.	0.3937	Ampere-turns per cm
Atmospheres	76.0	Cm of mercury
Atmospheres	29.92	Inches of mercury
Atmospheres	33.90	Feet of water
Atmospheres	14.70	Pounds per sq in.
British thermal units	252.0	Calories
British thermal units	778.2	Foot-pounds
British thermal units	3.960×10^{-4}	Horsepower-hours
British thermal units	0.2520	Kilogram-calories
British thermal units	107.6	Kilogram-meters
British thermal units	2.931×10^{-4}	Kilowatt-hours
British thermal units	1,055	Watt-seconds
B.t.u. per hour	2.931×10^{-4}	Kilowatts
B.t.u. per minute	2.359×10^{-2}	Horsepower
B.t.u. per minute	1.759×10^{-2}	Kilowatts
Bushels	1.244	Cubic feet
Centimeters	0.3937	Inches
Circular mils	5.067×10^{-6}	Square centimeters
Circular mils	0.7854×10^{-6}	Square inches

COMMONLY USED CONVERSION FACTORS (cont.)

Multiply	By	To Obtain
Circular mils	0.7854	Square mils
Cords	128	Cubic feet
Cubic centimeters	6.102×10^{-6}	Cubic inches
Cubic feet	0.02832	Cubic meters
Cubic feet	7.481	Gallons
Cubic feet	28.32	Liters
Cubic inches	16.39	Cubic centimeters
Cubic meters	35.31	Cubic feet
Cubic meters	1.308	Cubic yards
Cubic yards	0.7646	Cubic meters
Degrees (angle)	0.01745	Radians
Dynes	2.248×10^{-6}	Pounds
Ergs	1	Dyne-centimeters
Ergs	7.37×10^{-6}	Foot-pounds
Ergs	10^{-7}	Joules
Farads	10^6	Microfarads
Fathoms	6	Feet
Feet	30.48	Centimeters
Feet of water	.08826	Inches of mercury
Feet of water	304.8	Kg per square meter
Feet of water	62.43	Pounds per square ft.
Feet of water	0.4335	Pounds per square in.
Foot-pounds	1.285×10^{-2}	British thermal units
Foot-pounds	5.050×10^{-7}	Horsepower-hours
Foot-pounds	1.356	Joules
Foot-pounds	0.1383	Kilogram-meters
Foot-pounds	3.766×10^{-7}	Kilowatt-hours
Gallons	0.1337	Cubic feet
Gallons	231	Cubic inches
Gallons	3.785×10^{-3}	Cubic meters
Gallons	3.785	Liters
Gallons per minute	2.228×10^{-3}	Cubic feet per sec.
Gausses	6.452	Lines per square in.
Gilberts	0.7958	Ampere-turns
Henries	10^3	Millihenries
Horsepower	42.41	Btu per min.
Horsepower	2,544	Btu per hour

COMMONLY USED CONVERSION FACTORS (cont.)

Multiply	By	To Obtain
Horsepower	550	Foot-pounds per sec.
Horsepower	33,000	Foot-pounds per min.
Horsepower	1.014	Horsepower (metric)
Horsepower	10.70	Kg calories per min.
Horsepower	0.7457	Kilowatts
Horsepower (boiler)	33,520	Btu per hour
Horsepower-hours	2,544	British thermal units
Horsepower-hours	1.98×10^6	Foot-pounds
Horsepower-hours	2.737×10^5	Kilogram-meters
Horsepower-hours	0.7457	Kilowatt-hours
Inches	2.540	Centimeters
Inches of mercury	1.133	Feet of water
Inches of mercury	70.73	Pounds per square ft.
Inches of mercury	0.4912	Pounds per square in.
Inches of water	25.40	Kg per square meter
Inches of water	0.5781	Ounces per square in.
Inches of water	5.204	Pounds per square ft.
Joules	9.478×10^{-4}	British thermal units
Joules	0.2388	Calories
Joules	10^7	Ergs
Joules	0.7376	Foot-pounds
Joules	2.778×10^{-7}	Kilowatt-hours
Joules	0.1020	Kilogram-meters
Joules	1	Watt-seconds
Kilograms	2.205	Pounds
Kilogram-calories	3.968	British thermal units
Kilogram meters	7.233	Foot-pounds
Kg per square meter	3.281×10^{-3}	Feet of water
Kg per square meter	0.2048	Pounds per square ft.
Kg per square meter	1.422×10^{-3}	Pounds per square in.
Kilolines	10^3	Maxwells
Kilometers	3.281	Feet
Kilometers	0.6214	Miles
Kilowatts	56.87	Btu per min.
Kilowatts	737.6	Foot-pounds per sec.
Kilowatts	1.341	Horsepower
Kilowatts-hours	3409.5	British thermal units

COMMONLY USED CONVERSION FACTORS (cont.)

Multiply	By	To Obtain
Kilowatts-hours	2.655×10^6	Foot-pounds
Knots	1.152	Miles
Liters	0.03531	Cubic feet
Liters	61.02	Cubic inches
Liters	0.2642	Gallons
Log N_e or in N	0.4343	Log_{10} N
Log N	2.303	Log_e N or in N
Lumens per square ft.	1	Footcandles
Maxwells	10^{-3}	Kilolines
Megalines	10^6	Maxwells
Megaohms	10^6	Ohms
Meters	3.281	Feet
Meters	39.37	Inches
Meter-kilograms	7.233	Pound-feet
Microfarads	10^{-6}	Farads
Microhms	10^{-6}	Ohms
Microhms per cm cube	0.3937	Microhms per in. cube
Microhms per cm cube	6.015	Ohms per mil foot
Miles	5,280	Feet
Miles	1.609	Kilometers
Miner's inches	1.5	Cubic feet per min.
Ohms	10^{-6}	Megohms
Ohms	10^6	Microhms
Ohms per mil foot	0.1662	Microhms per cm cube
Ohms per mil foot	0.06524	Microhms per in. cube
Poundals	0.03108	Pounds
Pounds	32.17	Poundals
Pound-feet	0.1383	Meter-Kilograms
Pounds of water	0.01602	Cubic feet
Pounds of water	0.1198	Gallons
Pounds per cubic foot	16.02	Kg per cubic meter
Pounds per cubic foot	5.787×10^{-4}	Pounds per cubic in.
Pounds per cubic inch	27.68	Grams per cubic cm
Pounds per cubic inch	2.768×10^{-4}	Kg per cubic meter
Pounds per cubic inch	1.728	Pounds per cubic ft.
Pounds per square foot	0.01602	Feet of water
Pounds per square foot	4.882	Kg per square meter

COMMONLY USED CONVERSION FACTORS (cont.)

Multiply	By	To Obtain
Pounds per square foot	6.944×10^{-3}	Pounds per sq. in.
Pounds per square inch	2.307	Feet of water
Pounds per square inch	2.036	Inches of mercury
Pounds per square inch	703.1	Kg per square meter
Radians	57.30	Degrees
Square centimeters	1.973×10^{5}	Circular mils
Square feet	2.296×10^{-5}	Acres
Square feet	0.09290	Square meters
Square inches	1.273×10^{6}	Circular mils
Square inches	6.452	Square centimeters
Square kilometers	0.3861	Square miles
Square meters	10.76	Square feet
Square miles	640	Acres
Square miles	2.590	Square kilometers
Square millimeters	1.973×10^{3}	Circular mils
Square mils	1.273	Circular mils
Tons (long)	2,240	Pounds
Tons (metric)	2,205	Pounds
Tons (short)	2,000	Pounds
Watts	0.05686	Btu per minute
Watts	10^{7}	Ergs per sec.
Watts	44.26	Foot-pounds per min.
Watts	1.341×10^{-3}	Horsepower
Watts	14.34	Calories per min.
Watts-hours	3.412	British thermal units
Watts-hours	2,655	Footpounds
Watts-hours	1.341×10^{-3}	Horsepower-hours
Watts-hours	0.8605	Kilogram-calories
Watts-hours	376.1	Kilogram-meters
Webers	10^{8}	Maxwells

ELECTRICAL PREFIXES

Prefixes

Prefixes are used to avoid long expressions of units that are smaller and larger than the base unit. See Common Prefixes. For example, sentences 1 and 2 do not use prefixes. Sentences 3 and 4 use prefixes.

1. A solid-state device draws 0.000001 amperes (A).
2. A generator produces 100,000 watts (W).
3. A solid-state device draws 1 microampere (μA).
4. A generator produces 100 kilowatts (kW).

Converting Units

To convert between different units, the decimal point is moved to the left or right, depending on the unit. See Conversion Table. For example, an electronic circuit has a current flow of .000001 A. The current value is converted to simplest terms by moving the decimal point six places to the right to obtain 1.0μA (from Conversion Table).

$$.000001. \, A = 1.0 \, \mu A$$

Move decimal point
6 places to right

Common Electrical Quantities

Abbreviations are used to simplify the expression of common electrical quantities. See Common Electrical Quantities. For example, milliwatt is abbreviated mW, kilovolt is abbreviated kV, and ampere is abbreviated A.

COMMON PREFIXES

Symbol	Prefix	Equivalent
G	giga	1,000,000,000
M	mega	1,000,000
k	kilo	1000
base unit	—	1
m	milli	.001
u	micro	.000001
n	nano	.000000001

COMMON ELECTRICAL QUANTITIES

Variable	Name	Unit of Measure and Abbreviation
E	voltage	volt - V
I	current	ampere - A
R	resistance	ohm - Ω
P	power	watt - W
P	power (apparent)	volt-amp - VA
C	capacitance	farad - F
L	inductance	henry - H
Z	impedance	ohm - Ω
G	conductance	siemens - S
f	frequency	hertz - Hz
T	period	second - s

CONVERSION TABLE

Initial Units	Final Units						
	giga	mega	kilo	base unit	milli	micro	nano
giga	—	3R	6R	9R	12R	15R	18R
mega	3L	—	3R	6R	9R	12R	15R
kilo	6L	3L	—	3R	6R	9R	12R
base unit	9L	6L	3L	—	3R	6R	9R
milli	12L	9L	6L	3L	—	3R	6R
micro	15L	12L	9L	6L	3L	—	3R
nano	18L	15L	12L	9L	6L	3L	—

ELECTRICAL ABBREVIATIONS

Abbrev.	Term	Abbrev.	Term
A	Amps; armature; anode; ammeter	K	Kilo; cathode
Ag	Silver	L	Line; load
ALM	Alarm	LB-FT	Pounds per feet
AM	Ammeter	LB-IN	Pounds per inch
ARM	Armature	LRC	Locked rotor current
Au	Gold	M	Motor; motor starter contacts
BK	Black	MED	Medium
BL	Blue	N	Nirth
BR	Brown	NC	Normally closed
C	Celsius; centigrade	NO	Normally opened
CAP	Capacitor	NTDF	Nontime-delay fuse
CB	Circuit breaker	O	Orange
CCW	Counterclockwise	OCPD	Overcurrent protection device
CONT	Continuous	OL	Overloads
CPS	Cycles per second	OZ/IN	Ounces per inch
CR	Control relay	P	Power consumed
CT	Current transformer	PSI	Pounds per square inch
CW	Clockwise	PUT	Pull-up torque
D	Diameter	R	Resistance; radius; red; reverse
DP	Double-pole	REV	Reverse
DPDT	Double-pole, double-throw	RPM	Revolutions per minute
EMF	Electromotive force	S	Switch; series; slow; south
F	Fahrenheit; forward; fast	SCR	Silicon controlled rectifier
F	Field; forward	SF	Service factor
FLC	Full-load current	SP	Single-pole
FLT	Full-load torque	SPDT	Single-pole; double-throw
FREQ	Frequency	SPST	Single-pole; single-throw
FS	Float switch	SW	Switch
FTS	Foot switch	T	Terminal; torque
FWD	Forward	TD	Time delay
G	Green; gate	TDF	Time-delay fuse
GEN	Generator	TEMP	Temperature
GY	Gray	V	Volts; violet
H	Transformer, primary side	VA	Voltamps
HP	Horsepower	VAC	Volts alternating current
I	Current	VDC	Volts direct current
IC	Integrated circuit	W	White; watt
INT	Intermediate; interrupt	W/	With
ITB	Inverse time breaker	X	Transformer secondary side
ITCB	Instantaneous trip circuit breaker	Y	Yellow

CONVERSION TABLE FOR TEMPERATURE – °F/°C

°F	°C	°F	°C	°F	°C	°F	°C	°F	°C
-459.4	-273	-22.0	-30	35.6	.2	93.2	.34	150.8	.66
-418.0	-250	-18.4	-28	39.2	.4	96	.36	154.4	.68
-328.0	-200	-14.8	-26	42.8	.6	100.4	.38	158.0	.70
-238.0	-150	-11.2	-24	46.4	.8	104.0	.40	161.6	.72
-193.0	-125	-7.6	-22	50.0	.10	107.6	.42	165.2	.74
-148.0	-100	-4.0	-20	53.6	.12	111.2	.44	168.8	.76
-130.0	-90	-0.4	-18	57.2	.14	114.8	.46	172.4	.78
-112.0	-80	3.2	-16	60.8	.16	118.4	.48	176.0	.80
-94.0	-70	6.8	-14	64.4	.18	122.0	.50	179.6	.82
-76.0	-60	10.4	-12	68.0	.20	125.6	.52	183.2	.84
-58.0	-50	14.0	-10	71.6	.22	129.2	.54	186.8	.86
-40.0	-40	17.6	-8	75.2	.24	132.8	.56	190.4	.88
-36.4	-38	21.2	-6	78.8	.26	136.4	.58	194.0	.90
-32.8	-36	24.8	-4	82.4	.28	140.0	.60	197.6	.92
-29.2	-34	28.4	-2	86.0	.30	143.6	.62	201.2	.94
-25.6	-32	32.0	.0	89.6	.32	147.2	.64	204.8	.96

1 degree F is 1/180 of the difference between the temperature of melting ice and boiling water.
1 degree C is 1/100 of the difference between the temperature of melting ice and boiling water.

Absolute Zero = 273.16°C = -459.69°F

CONVERSION TABLE FOR TEMPERATURE – °F/°C (cont.)

°F	°C	°F	°C	°F	°C	°F	°C	°F	°C
208.4	.98	347.0	.175	590	.310	1004	.540	6332	.3500
212.0	.100	356.0	.180	608	.320	1040	.560	7232	.4000
221.0	.105	365.0	.185	626	.330	1076	.580	4500	.8132
230.0	.110	374.0	.190	644	.340	1112	.600	9032	.5000
239.0	.115	383.0	.195	662	.350	1202	.650	9932	.5500
248.0	.120	392.0	.200	680	.360	1292	.700	10832	.6000
257.0	.125	410.0	.210	698	.370	1382	.750	11732	.6500
266.0	.130	428.0	.220	716	.380	1472	.800	12632	.7000
275.0	.135	446.0	.230	734	.390	1562	.850	13532	.7500
284.0	.140	464.0	.240	752	.400	1652	.900	14432	.8000
293.0	.145	482.0	.250	788	.420	1742	.950	15332	.8500
302.0	.150	500.0	.260	824	.440	1832	.1000	16232	.9000
311.0	.155	518.0	.270	860	.460	2732	.1500	17132	.9500
320.0	.160	536.0	.280	896	.480	3632	.2000	18032	.10000
329.0	.165	554.0	.290	932	.500	4532	.2500		
338.0	.170	572.0	.300	968	.520	5432	.3000		

1 degree F is 1/180 of the difference between the temperature of melting ice and boiling water.
1 degree C is 1/100 of the difference between the temperature of melting ice and boiling water.

Absolute Zero = 273.16° C = -459.69° F

DECIBEL LEVELS OF SOUNDS

The definition of sound intensity is energy (erg) transmitted per 1 second over a square centimeter surface. Sounds are measured in decibels. A decibel (dB) change of 1 is the smallest change detected by humans.

Hearing Intensity	Decibel Level	Examples of Sounds
Barely Audible	0	Dead silence
		Audible hearing threshold
	10	Room (sound proof)
(Very Light)	20	Empty auditorium
		Ticking of a stopwatch
		Soft whispering
Audible	30	People talking quietly
Light	40	Quiet street noise without autos
Medium	45	Telephone operator
Loud	50	Fax machine in office
	60	Close conversation
Loud	70	Stereo system
		Computer printer
	80	Fire truck/Ambulance siren
		Cat/dog fight
Extremely Loud	90	Industrial machinery
		High school marching band
Damage Possible	100	Heavy duty grinder in a machine/welding shop
Damaging	100+	Begins ear damage
	110	Diesel engine of a train
	120	Lighting strike (thunderstorm)
		60 ton metal forming factory press
	130	60" fan in a bus vacuum system
	140	Commercial/Military jet engine
Ear Drum Shattering	194	Space shuttle engines
	225	16" Guns on a battleship

SOUND AWARENESS AND SAFETY

Sound Awareness Changes

The typical range of human hearing is 30 hertz - 15,000 hertz. Human hearing recognizes an increase of 20 decibels, such as a stereo sound level increase, as being four times as loud at the higher level than it was at the lower level.

Awareness in Human Hearing	Decibel Change
Noticeably Louder	10
Easily Audible	5
Faintly Audible	3

HEARING PROTECTION LEVELS

Because of the occupational safety and health act of 1970, hearing protection is mandatory if the following time exposures to decibel levels are exceeded because of possible damage to human hearing.

Decibel Level	Time Exposure Per Day
115	15 minutes
110	30 minutes
105	1 hour
102	1-1/2 hours
100	2 hours
97	3 hours
95	4 hours
92	6 hours
90	8 hours

TRIGONOMETRIC FORMULAS — RIGHT TRIANGLE

Angles = X, Y, Z
Distances = x, y, z
Area = $\dfrac{x \, y}{2}$

Pythagorean Theorem states

That $x^2 + y^2 = z^2$

Thus $x = \sqrt{z^2 - y^2}$

Thus $y = \sqrt{z^2 - x^2}$

Thus $z = \sqrt{x^2 + y^2}$

$\sin X = \dfrac{x}{z}$ $\cos X = \dfrac{y}{z}$

$\tan X = \dfrac{x}{y}$ $\cot X = \dfrac{y}{x}$

Given x and z, find X, Y and y

$\sin X = \dfrac{x}{z} = \cos Y,\ y = \sqrt{(z^2 - x^2)} = z \sqrt{1 - \dfrac{x^2}{z^2}}$

Given x and y, find X, Y and z

$\tan X = \dfrac{x}{y} = \cot Y,\ z = \sqrt{x^2 + y^2} = x \sqrt{1 + \dfrac{y^2}{x^2}}$

Given X and x, find Y, y and z

$Y = 90^\circ - X,\ y = x \cot X,\ z = \dfrac{x}{\sin X}$

Given X and z, find Y, x and y

$Y = 90^\circ - X,\ x = z \sin X,\ y = z \cos X$

Given X and z, find Y, x and z

$Y = 90^\circ - X,\ x = y \tan X,\ z = \dfrac{y}{\cos X}$

TRIGONOMETRIC FORMULAS — OBLIQUE TRIANGLES

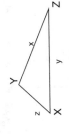

Given x, y and z, Find X, Y and Z

$$s = \frac{x+y+z}{2}, \quad \sin\frac{1}{2}X = \sqrt{\frac{(s-y)(s-z)}{yz}}$$

$$\sin\frac{1}{2}Y = \sqrt{\frac{(s-x)(s-z)}{xz}}, \quad C = 180° - (X+Y)$$

Given x, y and z, find the Area

$$s = \frac{x+y+z}{2}, \quad \text{Area} = \sqrt{S(s-x)(s-y)(s-z)}$$

$$\text{Area} = \frac{yz \sin X}{2}, \quad \text{Area} = \frac{x^2 \sin Y \sin Z}{2 \sin X}$$

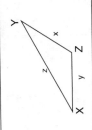

Given x, Y and Z, find X, Y and z

$$X + Y = 180° - Z, \quad z = \frac{x \sin Z}{\sin X}, \quad \tan X = \frac{x \sin Z}{y - (x \cos Z)}$$

Given X, x and y, Find Y, Z and z

$$\sin Y = \frac{y \sin X}{x}, \quad Z = 180° - (X+Y), \quad z = \frac{x \sin Z}{\sin X}$$

Given X, Y and x, Find y, Z and z

$$y = \frac{x \sin Y}{\sin X}, \quad Z = 180° - (X+Y), \quad z = \frac{x \sin Z}{\sin X}$$

10-13

TRIGONOMETRIC FORMULAS — SHAPES

Equilateral Triangle

X = Sides (Equal Lengths)

$Area = X^2 \sqrt{\dfrac{3}{4}} = .433\ X^2$

$Perimeter = 3\ X$

$H = \dfrac{X}{2} \sqrt{3} = .866\ X$

Annulus

C_1 and R_1 = Inside Circle

C_2 and R_2 = Outside Circle

C = Circumference

R = Radius

$Area = \pi\ (R_1 + R_2)\ (R_2 - R_1)$

$Area = \left((C_2)^2 - (C_1)^2 \right) .7854$

Trapezium

Perimeter is the

Sum of L, M, N and O

$Area = \dfrac{(S + T)\ Q + RS + PT}{2}$

TRIGONOMETRIC FORMULAS — SHAPES (cont.)

Trapezoid

Perimeter =
The Sum of the
lengths of all
four sides

$Area = \dfrac{(X + Y)}{2}$

Quadrilateral

$Area = \dfrac{L_1 \cdot L_2 \cdot \sin \theta}{2}$

Where θ =
Degrees
of Angle

Rectangle

$Area = XY$

Diagonal Line (D)

$= \sqrt{X^2 + Y^2}$

Perimeter =
$2(X + Y)$

If a square
then $X = Y$

Parallelogram

Where θ =
Degrees
of Angle

$Area =$
$XH = XY \sin \theta$
Perimeter =
$2(X + Y)$

10-15

DECIMAL EQUIVALENTS OF FRACTIONS

8ths	32nds	64ths	64ths
1/8 = .125	1/32 = .03125	1/64 = 0.15625	33/64 = .515625
1/4 = .250	3/32 = .09375	3/64 = .046875	35/64 = .546875
3/8 = .375	5/32 = .15625	5/64 = .078125	37/64 = .57812
1/2 = .500	7/32 = .21875	7/64 = .109375	39/64 = .609375
5/8 = .625	9/32 = .28125	9/64 = .140625	41/64 = .640625
3/4 = .750	11/32 = .34375	11/64 = .171875	43/64 = .671875
7/8 = .875	13/32 = .40625	13/64 = .203128	45/64 = .703125
16ths	15/32 = .46875	15/64 = .234375	47/64 = .734375
1/16 = .0625	17/32 = .53125	17/64 = .265625	49/64 = .765625
3/16 = .1875	19/32 = .59375	19/64 = .296875	51/64 = 3796875
5/16 = .3125	21/32 = .65625	21/64 = .328125	53/64 = .828125
7/16 = .4375	23/32 = .71875	23/64 = .359375	55/64 = .859375
9/16 = .5625	25/32 = .78125	25/64 = .390625	57/64 = .890625
11/16 = .6875	27/32 = .84375	27/64 = .421875	59/64 = .921875
13/16 = .8125	29/32 = .90625	29/64 = .453125	61/64 = .953125
15/16 = .9375	31/32 = .96875	31/64 = .484375	63/64 = .984375

COMMON ENGINEERING UNITS AND THEIR RELATIONSHIP

Quantity	SI Metric Units/Symbols	Customary Units	Relationship of Units
Acceleration	meters per second squared (m/s^2)	feet per second squared (ft/s^2)	$m/s^2 = ft/s^2 \times 3.281$
Area	square meter (m^2) square millimeter (mm^2)	square foot (ft^2) square inch (in^2)	$m^2 = ft^2 \times 10.764$ $mm^2 = in^2 \times 0.00155$
Density	kilograms per cubic meter (kg/m^3) grams per cubic centimeter (g/cm^3)	pounds per cubic foot (lb/ft^3) pounds per cubic inch (lb/in^3)	$kg/m^3 = lb/ft^3 \times 16.02$ $g/cm^3 = lb/in^3 \times 0.036$
Work	Joule (J)	foot pound force (ft lbf or ft lb)	$J = ft\ lbf \times 1.356$
Heat	Joule (J)	British thermal unit (Btu) Calorie (Cal)	$J = Btu \times 1.055$ $J = cal \times 4.187$
Energy	kilowatt (kW)	Horsepower (HP)	$kW = HP \times 0.7457$
Force	Newton (N) Newton (N)	Pound-force (lbf, lb · f, or lb) kilogram-force (kgf, kg · f, or kp)	$N = lbf \times 4.448$ $N = \dfrac{kgf}{9.807}$
Length	meter (m) millimeter (mm)	foot (ft) inch (in)	$m = ft \times 3.281$ $mm = \dfrac{in}{25.4}$
Mass	kilogram (kg) gram (g)	pound (lb) ounce (oz)	$kg = lb \times 2.2$ $g = \dfrac{oz}{28.35}$
Stress	Pascal = Newton per second (Pa = N/s)	pounds per square inch (lb/in^2 or psi)	$Pa = lb/in^2 \times 6{,}895$
Temperature	degree Celsius (°C)	degree Fahrenheit (°F)	$°C = \dfrac{°F - 32}{1.8}$
Torque	Newton meter (N · m)	foot-pound (ft lb) inch-pound (in lb)	$N \cdot m = ft\ lbf \times 1.356$ $N \cdot m = in\ lbf \times 0.113$
Volume	cubic meter (m^3) cubic centimeter (cm^3)	cubic foot (ft^3) cubic inch (in^3)	$m^3 = ft^3 \times 35.314$ $cm^3 = \dfrac{in^3}{16.387}$

COMMONLY USED GEOMETRICAL RELATIONSHIPS

Diameter of a circle x 3.1416 = Circumference.

Radius of a circle x 6.283185 = Circumference.

Square of the radius of a circle x 3.1416 = Area.

Square of the diameter of a circle x 0.7854 = Area.

Square of the circumference of a circle x 0.07958 = Area.

Half the circumference of a circle x half its diameter = Area.

Circumference of a circle x 0.159155 = Radius.

Square root of the area of a circle x 0.56419 = Radius.

Circumference of a circle x 0.31831 = Diameter.

Square root of the area of a circle x 1.12838 = Diameter.

Diameter of a circle x 0.866 = Side of an inscribed equilateral triangle.

Diameter of a circle x 0.7071 = Side of an inscribed square.

Circumference of a circle x 0.225 = Side of an inscribed square.

Circumference of a circle x 0.282 = Side of an equal square.

Diameter of a circle x 0.8862 = Side of an equal square.

Base of a triangle x one-half the altitude = Area.

Multiplying both diameters and .7854 together = Area of an ellipse.

Surface of a sphere x one-sixth of its diameter = Volume.

Circumference of a sphere x its diameter = Surface.

Square of the diameter of a sphere x 3.1416 = Surface.

Square of the circumference of a sphere x 0.3183 = Surface.

Cube of the diameter of a sphere x 0.5236 = Volume.

Cube of the circumference of a sphere x 0.016887 = Volume.

Radius of a sphere x 1.1547 = Side of an inscribed cube.

Diameter of a sphere divided by $\sqrt{3}$ = Side of an inscribed cube.

Area of its base x one-third of its altitude = Volume of a cone or
pyramid whether round, square or triangular.

Area of one of its sides x 6 = Surface of the cube.

Altitude of trapezoid x one-half the sum of its parallel sides = Area.

CHAPTER 11
MATERIALS AND TOOLS

STRENGTH GAIN VS. PULL ANGLE

The weight bearing capacity of a strap, for example, increases by the factor K shown below as the angle of the strap decreases. A 100 lb. capacity strap at a 60° pulling angle loses 50 lbs. of weight bearing ability.

At perfectly vertical, the ability is 100%.

$D° = $ Pull Angle

Strap →

Weight

K	D°	K	D°	K	D°
.7412	75	.3572	50	.0937	25
.6580	70	.2929	45	.0603	20
.5774	65	.2340	40	.0341	15
.5000	60	.1340	35	.0152	10
.4264	55	.1208	30	.0038	5

LENGTH OF WIRE CABLE PER REEL

Cable length (feet) = X (X + Y) Z (K)
Use equation above to determine the length of wire cable that is wound smoothly on a reel. The dimensions for X, Y and Z are in inches.

(K)	Cable Dia. (Inches)	(K)	Cable Dia. (Inches)
.0476	2.250	.239	1.000
.0532	2.125	.308	.875
.0597	2.000	.428	.750
.0675	1.875	.607	.625
.0770	1.750	.741	.563
.0886	1.625	.925	.500
.107	1.500	1.19	.438
.127	1.375	1.58	.375
.152	1.250	2.21	.313
.191	1.125	3.29	.250

SHEET METAL SCREW CHARACTERISTICS

Screw Size #	Screw Dia. (Inches)	Diameter of Pierced Hole (Inches)	Hole Size #	Thickness of Metal – Gauge #
4	.112	.086	44	28
		.086	44	26
		.093	42	24
		.098	42	22
		.100	40	20
6	.138	.111	39	28
		.111	39	26
		.111	39	24
		.111	38	22
		.111	36	20
7	.155	.121	37	28
		.121	37	26
		.121	35	24
		.121	33	22
		.121	32	20
		–	31	18
8	.165	.137	33	26
		.137	33	24
		.137	32	22
		.137	31	20
		–	30	18
10	.191	.158	30	26
		.158	30	24
		.158	30	22
		.158	29	20
		.158	25	18
12	.218	–	26	24
		.185	25	22
		.185	24	20
		.185	22	18
14	.251	–	15	24
		.212	12	22
		.212	11	20
		.212	9	18

Deviations in materials and conditions could require variations from these dimensions.

STANDARD WOOD SCREW CHARACTERISTICS (INCHES)

Screw Size #	Wood Screw Standard Lengths (Inches)	Size of Pilot Hole		Size of Shank Hole	
		Softwood Bit #	Hardwood Bit #	Clearance Bit #	Hole Diameter (Inches)
0	1/4	75	66	52	.060
1	1/4 to 3/8	71	57	47	.073
2	1/4 to 1/2	65	54	42	.086
3	1/4 to 5/8	58	53	37	.099
4	3/8 to 3/4	55	51	32	.112
5	3/8 to 3/4	53	47	30	.125
6	3/8 to 1-1/2	52	44	27	.138
7	3/8 to 1-1/2	51	39	22	.151
8	1/2 to 2	48	35	18	.164
9	5/8 to 2-1/4	45	33	14	.177
10	5/8 to 2-1/2	43	31	10	.190
11	3/4 to 3	40	29	4	.203
12	7/8 to 3-1/2	38	25	2	.216
14	1 to 4-1/2	32	14	D	.242
16	1-1/4 to 5-1/2	29	10	I	.268
18	1-1/2 to 6	26	6	N	.294
20	1-3/4 to 6	19	3	P	.320
24	3-1/2 to 6	15	D	V	.372

ALLEN HEAD AND MACHINE SCREW
BOLT AND TORQUE CHARACTERISTICS

Number of Threads Per Inch	Allen Head And Mach. Screw Bolt Size	Allen Head Case H Steel 160,000 psi	Mach. Screw Yellow Brass 60,000 psi	Mach. Screw Silicone Bronze 70,000 psi
		Torque in Foot-Pounds or Inch-Pounds		
4.5	2"	8800	–	–
5	1-3/4"	6100	–	–
6	1-1/2"	3450	655	595
6	1-3/8"	2850	–	–
7	1-1/4"	2130	450	400
7	1-1/8"	1520	365	325
8	1"	970	250	215
9	7/8"	640	180	160
10	3/4"	400	117	104
11	5/8"	250	88	78
12	9/16"	180	53	49
13	1/2"	125	41	37
14	7/16"	84	30	27
16	3/8"	54	20	17
18	5/16"	33	125 in#	110 in#
20	1/4"	16	70 in#	65 in#
24	#10	60	22 in#	20 in#
32	#8	46	19 in#	16 in#
32	#6	21	10 in#	8 in#
40	#5	–	7.2 in#	6.4 in#
40	#4	–	4.9 in#	4.4 in#
48	#3	–	3.7 in#	3.3 in#
56	#2	–	2.3 in#	2 in#

For fine thread bolts, increase by 9%.

HEX HEAD BOLT AND TORQUE CHARACTERISTICS

BOLT MAKE-UP IS STEEL WITH COARSE THREADS

Number of Threads Per Inch	Hex Head Bolt Size (Inches)	⬡ SAE 0-1-2 74,000 psi	⬡ SAE Grade 3 100,000 psi	⬡ SAE Grade 5 120,000 psi
		Torque = Foot-Pounds		
4.5	2	2750	5427	4550
5	1-3/4	1900	3436	3150
6	1-1/2	1100	1943	1775
6	1-3/8	900	1624	1500
7	1-1/4	675	1211	1105
7	1-1/8	480	872	794
8	1	310	551	587
9	7/8	206	372	382
10	3/4	155	234	257
11	5/8	96	145	154
12	9/16	69	103	114
13	1/2	47	69	78
14	7/16	32	47	54
16	3/8	20	30	33
18	5/16	12	17	19
20	1/4	6	9	10

For fine thread bolts, increase by 9%.

HEX HEAD BOLT AND TORQUE CHARACTERISTICS (cont.)

BOLT MAKE-UP IS STEEL WITH COARSE THREADS

Number of Threads Per Inch	Hex Head Bolt Size (Inches)	SAE Grade 6 133,000 psi	SAE Grade 7 133,000 psi	SAE Grade 8 150,000 psi
		Torque = Foot-Pounds		
4.5	2	7491	7500	8200
5	1-3/4	5189	5300	5650
6	1-1/2	2913	3000	3200
6	1-3/8	2434	2500	2650
7	1-1/4	1815	1825	1975
7	1-1/8	1304	1325	1430
8	1	825	840	700
9	7/8	550	570	600
10	3/4	350	360	380
11	5/8	209	215	230
12	9/16	150	154	169
13	1/2	106	110	119
14	7/16	69	71	78
16	3/8	43	44	47
18	5/16	24	25	29
20	1/4	12.5	13	14

For fine thread bolts, increase by 9%.
For special alloy bolts, obtain torque rating from the manufacturer.

WHITWORTH HEX HEAD BOLT AND TORQUE CHARACTERISTICS

BOLT MAKE-UP IS STEEL WITH COARSE THREADS

Number of Threads Per Inch	Whitworth Type Hex Head Bolt Size (Inches)	Grades A & B 62,720 psi	Grade S 112,000 psi	Grade T 123,200 psi	Grade V 145,600 psi
			Torque = Foot-Pounds		
8	1	276	497	611	693
9	7/8	186	322	407	459
11	3/4	118	213	259	287
11	5/8	73	128	155	175
12	9/16	52	94	111	128
12	1/2	36	64	79	89
14	7/16	24	43	51	58
16	3/8	15	27	31	36
18	5/16	9	15	18	21
20	1/4	5	7	9	10

For fine thread bolts, increase by 9%.

11-7

METRIC HEX HEAD BOLT AND TORQUE CHARACTERISTICS

BOLT MAKE-UP IS STEEL WITH COARSE THREADS

(Metric Type) Thread Pitch	(Dimensions in Millimeters) Bolt Size	5D Standard 5D 71,160 psi	8G Standard 8G 113,800 psi	10K Standard 10K 142,000 psi	12K Standard 12K 170,674 psi
		Torque = Foot-Pounds			
3.0	24	261	419	570	689
2.5	22	182	284	394	464
2.0	18	111	182	236	183
2.0	16	83	132	175	208
1.25	14	55	89	117	137
1.25	12	34	54	70	86
1.25	10	19	31	40	49
1.0	8	10	16	22	27
1.0	6	5	6	8	10

For fine thread bolts, increase by 9%.

11-8

TIGHTENING TORQUE IN POUND-FEET-SCREW FIT

Wire Size, AWG/kcmil	Driver	Bolt	Other
18-16	1.67	6.25	4.2
14-8	1.67	6.25	6.125
6-4	3.0	12.5	8.0
3-1	3.2	21.00	10.40
0-2/0	4.22	29	12.5
3/0-200	–	37.5	17.0
250-300	–	50.0	21.0
400	–	62.5	21.0
500	–	62.5	25.0
600-750	–	75.0	25.0
800-1000	–	83.25	33.0
1250-2000	–	83.26	42.0

SCREW TORQUES

Screw Size, Inches Across, Hex Flats	Torque, Pound-Feet
1/8	4.2
5/32	8.3
3/16	15
7/32	23.25
1/4	42

STANDARD TAPS AND DIES (IN INCHES)

Thread Size	Coarse			Fine		
	Drill Size	Threads Per Inch	Decimal Size	Drill Size	Threads Per Inch	Decimal Size
4"	3	4	3.75			
3-3/4"	3	4	3.5			
3-1/2"	3	4	3.25			
3-1/4"	3	4	3.0			
3"	2	4	2.75			
2-3/4"	2	4	2.5			
2-1/2"	2	4	2.25			
2-1/4"	2	4.5	2.0313			
2"	1	4.5	1.7813			
1-3/4"	1	2	1.5469			
1-1/2"	1	6	1.3281	1-27/64"	12	1.4219
1-3/8"	1	6	1.2188	1-19/64"	12	1.2969
1-1/4"	1	7	1.1094	1-11/64"	12	1.1719
1-1/8"	63/64"	7	.9844	1-3/64"	12	1.0469
1"	7/8"	8	.8750	15/16"	14	.9375
7/8"	49/64"	9	.7656	13/16"	14	.8125
3/4"	21/32"	10	.6563	11/16"	16	.6875
5/8"	17/32"	11	.5313	37/64"	18	.5781
9/16"	31/64"	12	.4844	33/64"	18	.5156
1/2"	27/64"	13	.4219	29/64"	20	.4531
7/16"	U	14	.368	25/64"	20	.3906
3/8"	5/16"	16	.3125	Q	24	.332
5/16"	F	18	.2570	I	24	.272
1/4"	#7	20	.201	#3	28	.213
#12	#16	24	.177	#14	28	.182
#10	#25	24	.1495	#21	32	.159
3/16"	#26	24	.147	#22	32	.157
#8	#29	32	.136	#29	36	.136
#6	#36	32	.1065	#33	40	.113
#5	#38	40	.1015	#37	44	.104
1/8"	3/32"	32	.0938	#38	40	.1015
#4	#43	40	.089	#42	48	.0935
#3	#47	48	.0785	#45	56	.082
#2	#50	56	.07	#50	64	.07
#1	#53	64	.0595	#53	72	.0595
#0	–	–	–	3/64"	80	.0469

(mm) Thread Pitch	TAPS AND DIES — METRIC CONVERSIONS			
	Fine Thread Size		Tap Drill Size	
	Inches	mm	Inches	mm
4.5	1.6535	42	1.4567	37.0
4.0	1.5748	40	1.4173	36.0
4.0	1.5354	39	1.3779	35.0
4.0	1.4961	38	1.3386	34.0
4.0	1.4173	36	1.2598	32.0
3.5	1.3386	34	1.2008	30.5
3.5	1.2992	33	1.1614	29.5
3.5	1.2598	32	1.1220	28.5
3.5	1.1811	30	1.0433	26.5
3.0	1.1024	28	.9842	25.0
3.0	1.0630	27	.9449	24.0
3.0	1.0236	26	.9055	23.0
3.0	.9449	24	.8268	21.0
2.5	.8771	22	.7677	19.5
2.5	.7974	20	.6890	17.5
2.5	.7087	18	.6102	15.5
2.0	.6299	16	.5118	14.0
2.0	.5512	14	.4724	12.0
1.75	.4624	12	.4134	10.5
1.50	.4624	12	.4134	10.5
1.50	.3937	11	.3780	9.6
1.50	.3937	10	.3386	8.6
1.25	.3543	9	.3071	7.8
1.25	.3150	8	.2677	6.8
1.0	.2856	7	.2362	6.0
1.0	.2362	6	.1968	5.0
.90	.2165	5.5	.1811	4.6
.80	.1968	5	.1653	4.2
.75	.1772	4.5	.1476	3.75
.70	.1575	4	.1299	3.3
.75	.1575	4	.1279	3.25
.60	.1378	3.5	.1142	2.9
.60	.1181	3	.0945	2.4
.50	.1181	3	.0984	2.5
.45	.1124	2.6	.0827	2.1
.45	.0984	2.5	.0787	2.0
.40	.0895	2.3	.0748	1.9
.40	.0787	2	.0630	1.6
.45	.0787	2	.0590	1.5
.35	.0590	1.5	.0433	1.1

RECOMMENDED DRILLING SPEEDS (RPMS)

Material	Bit Sizes	RPM Speed Range	
Glass	Special Metal Tube Drilling	700	
Plastics	7/16" and larger	500	– 1000
	3/8"	1500	– 2000
	5/16"	2000	– 2500
	1/4"	3000	– 3500
	3/16"	3500	– 4000
	1/8"	5000	– 6000
	1/16" and smaller	6000	– 6500
Woods	1" and larger	700	– 2000
	3/4" to 1"	2000	– 2300
	1/2" to 3/4"	2300	– 3100
	1/4" to 1/2"	3100	– 3800
	1/4" and smaller	3800	– 4000
	carving / routing	4000	– 6000
Soft Metals	7/16" and larger	1500	– 2500
	3/8"	3000	– 3500
	5/16"	3500	– 4000
	1/4"	4500	– 5000
	3/16"	5000	– 6000
	1/8"	6000	– 6500
	1/16" and smaller	6000	– 6500
Steel	7/16" and larger	500	– 1000
	3/8"	1000	– 1500
	5/16"	1000	– 1500
	1/4"	1500	– 2000
	3/16"	2000	– 2500
	1/8"	3000	– 4000
	1/16" and smaller	5000	– 6500
Cast Iron	7/16" and larger	1000	– 1500
	3/8"	1500	– 2000
	5/16"	1500	– 2000
	1/4"	2000	– 2500
	3/16"	2500	– 3000
	1/8"	3500	– 4500
	1/16" and smaller	6000	– 6500

TORQUE LUBRICATION EFFECTS IN FOOT-POUNDS

Lubricant	5/16" - 18 Thread	1/2" - 13 Thread	Torque Decrease
Graphite	13	62	49 - 55%
Mily Film	14	66	45 - 52%
White Grease	16	79	35 - 45%
Sae 30	16	79	35 - 45%
Sae 40	17	83	31 - 41%
Sae 20	18	87	28 - 38%
Plated	19	90	26 - 34%
No Lube	29	121	0%

METALWORKING LUBRICANTS

Materials	Threading	Lathing	Drilling
Machine Steels	Dissolvable Oil Mineral Oil Lard Oil	Dissolvable Oil	Dissolvable Oil Sulpherized Oil Min. Lard Oil
Tool Steels	Lard Oil Sulpherized Oil	Dissolvable Oil	Dissolvable Oil Sulpherized Oil
Cast Irons	Sulpherized Oil Dry Min. Lard Oil	Dissolvable Oil Dry	Dissolvable Oil Dry Air Jet
Malleable Irons	Soda Water Lard Oil	Soda Water Dissolvable Oil	Soda Water Dry
Aluminums	Kerosene Dissolvable Oil Lard Oil	Dissolvable Oil	Kerosene Dissolvable Oil
Brasses	Dissolvable Oil Lard Oil	Dissolvable Oil	Kerosene Dissolvable Oil Dry
Bronzes	Dissolvable Oil Lard Oil	Dissolvable Oil	Dissolvable Oil Dry
Coppers	Dissolvable Oil Lard Oil	Dissolvable Oil	Kerosene Dissolvable Oil Dry

TYPES OF SOLDERING FLUX

To Solder	Use
For cast iron	Cuprous oxide
For galvanized iron, galvanized, steel, tin, zinc	Hydrochloric acid
For pewter and lead	Organic
For brass, copper, gold, iron, silver, steel	Borax
For brass, bronze, cadmium, copper, lead, silver	Resin
For brass, copper, gun metal, iron, nickel, tin, zinc	Ammonia chloride
For bismuth, brass, copper, gold, silver, tin	Zinc chloride
For silver	Sterling
For pewter and lead	Tallow
For stainless only	Stainless steel (only)

HARD SOLDER ALLOYS

To hard solder	Copper %	Gold %	Silver %	Zinc %
Gold	22	67	11	
Silver	20		70	10
Hard brass	45			55
Soft brass	22			78
Copper	50			50
Cast iron	55			45
Steel and iron	64			36

SOFT SOLDER ALLOYS

To soft solder	Lead %	Tin %	Zinc %	Bism %	Other %
Gold	33	67			
Silver	33	67			
Brass	34	66			
Copper	40	60			
Steel and iron	50	50			
Galvanized steel	42	58			
Tinned steel	36	64			
Zinc	45	55			
Block Tin	1	99			
Lead	67	33			
Gun metal	37	63			
Pewter	25	25		50	
Bismuth	33	33		34	
Aluminum		70	25		5

PROPERTIES OF WELDING GASES

Type of Gas	Characteristics	Common Tank Sizes (cu. ft.)
Acetylene	C_2H_2, explosive gas, flammable, garlic - like odor, colorless, dangerous if used in pressures over 15 psig (30 psig absolute)	10, 40, 75 100, 300
Argon	Ar, non-explosive inert gas, tasteless, odorless, colorless	131, 330 4754 (Liquid)
Carbon Dioxide	CO_2, Non-explosive inert gas, tasteless, odorless, colorless (in large quantities is toxic)	20 lbs., 50 lbs.
Helium	He, Non-explosive inert gas, tasteless, odorless, colorless	221
Hydrogen	H2, explosive gas, tasteless, odorless, colorless	191
Nitrogen	N2, Non-explosive inert gas, tasteless, odorless, colorless	20, 40, 80 113, 225
Oxygen	O2, Non-explosive gas, tasteless, odorless, colorless, supports combustion	20, 40, 80 122, 244 4500 (liquid)

WELDING RODS – 36" LONG

Rod (Dia.) Size	Number of Rods Per Pound			
	Aluminum	Brass	Cast Iron	Steel
3/8"	-	1.0	.25	1.0
5/16"	-	-	.50	1.33
1/4"	6.0	2.0	2.25	2.0
3/16"	9.0	3.0	5.50	3.5
5/32"	-	-	-	5.0
1/8"	23.0	7.0	-	8.0
3/32"	41.0	13.0	-	14.0
1/16"	91.0	29.0	-	31.0

CABLE CLAMPS PER WIRE ROPE SIZE

Wire Rope Diameter (Inches)	# of Clamps Required	Clip Spacing (Inches)	Rope Turn-back (Inches)
1/8	2	3	3-1/4
3/16	2	3	3-3/4
1/4	2	3-1/4	4-3/4
5/16	2	3-1/4	5-1/4
3/8	2	4	6-1/2
7/16	2	4-1/2	4
1/2	3	5	11-1/2
9/16	3	5-1/2	12
5/8	3	5-3/4	12
3/4	4	6-3/4	18
7/8	4	8	19
1	5	8-3/4	26
1-1/8	6	9-3/4	34
1-1/4	6	10-3/4	37
1-7/16	7	11-1/2	44
1-1/2	7	12-1/2	48
1-5/8	7	13-1/4	51
1-3/4	7	14-1/2	53
2	8	16-1/2	71
2-1/4	8	16-1/2	73
2-1/2	9	17-3/4	84
2-3/4	10	18	100
3	10	18	106

TYPES OF FIRE EXTINGUISHERS

Today they are virtually standard equipment in a business or residence and are rated by the make up of the fire they will extinguish.

TYPE A: To extinguish fires involving trash, cloth, paper and other wood or pulp based materials. The flames are put out by water based ingredients or dry chemicals.

TYPE B: To extinguish fires involving greases, paints, solvents, gas and other petroleum based liquids. The flames are put out by cutting off oxygen and stopping the release of flammable vapors. Dry chemicals, foams and halon are used.

TYPE C: To extinguish fires involving electricity. The combustion is put out the same way as with a type B extinguisher, but, most importantly, the chemical in a type C <u>MUST</u> be non-conductive to electricity in order to be safe and effective.

TYPE D: To extinguish fires involving combustible metals. Please be advised to obtain important information from your local fire department on the requirements for type D fire extinguishers for your area.

Any combination of letters indicate that an extinguisher will put out more than one type of fire. A type BC will put out two types of fires. The size of the fire to be extinguished is shown by a number in front of the letter such as 100A.
The following formulas apply:

Class "1A": Will extinguish 25 burning sticks 40 inches long.

Class "1B": Will extinguish a paint thinner fire 2.5 square feet in size.

A 100B fire extinguisher will put out a fire 100 times larger than a type 1B.

Here are some basic guidelines to follow:

- By using a type ABC you will cover most basic fires.
- Use fire extinguishers with a gauge and ones that are constructed with metal. Also note if the unit is U.L. approved.
- Utilize more than one extinguisher and be sure that each unit is mounted in a clearly visible and accessible manner.
- After purchasing any fire extinguisher always review the basic instructions for its intended use. Never deviate from the manufacturers' guidelines. Following this simple procedure could end up saving lives.

PULLEYS AND GEARS

For single reduction or increase of speed by means of belting where the speed at which each shaft should run is known, and one pulley is in place:

Multiply the diameter of the pulley which you have by the number of revolutions per minute that its shaft makes; divide this product by the speed in revolutions per minute at which the second shaft should run. The result is the diameter of pulley to use.

Where both shafts with pulleys are in operation and the speed of one is known:

Multiply the speed of the shaft by diameter of its pulley and divide this product by diameter of pulley on the other shaft. The result is the speed of the second shaft.

Where a countershaft is used, to obtain size of main driving or driven pulley, or speed of main driving or driven shaft, it is necessary to calculate, as above, between the known end of the transmission and the countershaft, then repeat this calculation between the countershaft and the unknown end.

A set of gears of the same pitch transmits speeds in proportion to the number of teeth they contain. Count the number of teeth in the gear wheel and use this quantity instead of the diameter of pulley, mentioned above, to obtain number of teeth cut in unknown gear, or speed of second shaft.

FORMULAS FOR FINDING PULLEY SIZES

$$d = \frac{D \times S}{S'} \qquad D = \frac{d \times S'}{S}$$

d = diameter of driven pulley.

D = diameter of driving pulley.

S = number of revolutions per minute of driving pulley.

S' = number of revolutions per minute of driven pulley.

PULLEYS AND GEARS (cont.)

Formulas For Finding Gear Sizes:

$$n = \frac{N \times S}{S'} \qquad N = \frac{n \times S'}{S}$$

n = number of teeth in pinion (driving gear).

N = number of teeth in gear (driven gear).

S = number of revolutions per minute of pinion.

S' = number of revolutions per minute of gear.

Formulas To Determine Shaft Diameter:

The formula for determining the size of steel shaft for transmitting a given power at a given speed is as follows:

$$\text{diameter of shaft in inches} = \sqrt[3]{\frac{K \times HP}{RPM}}$$

when HP = the horse power to be transmitted

RPM = speed of shaft

K = factor which varies from 50 to 125 depending on type of shaft and distance between supporting bearings.

For line shaft having bearings 8 feet apart:

K = 90 for turned shafting

K = 70 for cold-rolled shafting

Formula To Determine Belt Length:

The following formula is used to determine the length of belting:

$$\text{length of belt} = \frac{3.14 (D + d)}{2} + 2 \sqrt{X^2 + \left(\frac{D - d}{2}\right)^2}$$

when D = diameter of large pulley

d = diameter of small pulley

X = distance between centers of shafting

STANDARD "V" BELT LENGTHS

A BELTS		
BELT # Standard	Pitch Length	Outside Length
A26	27.3	28.0
A31	32.3	33.0
A35	36.3	37.0
A38	39.3	40.0
A42	43.3	44.0
A46	47.3	48.0
A51	52.3	53.0
A55	56.3	57.0
A60	61.3	62.0
A68	69.3	70.0
A75	76.3	77.0
A80	81.3	82.0
A85	86.3	87.0
A90	91.3	92.0
A96	97.3	98.0
A105	106.3	107.0
A112	113.3	114.0
A120	121.3	122.0
A128	129.3	130.0

B BELTS		
BELT # Standard	Pitch Length	Outside Length
B35	36.8	38.0
B38	39.8	41.0
B42	43.8	45.0
B46	47.8	49.0
B51	52.8	54.0
B55	56.8	58.0
B60	61.8	63.0
B68	69.8	71.0
B75	76.8	78.0
B81	82.8	84.0
B85	86.8	88.0
B90	91.8	93.0
B97	98.8	100.0
B105	106.8	108.0
B112	113.8	115.0
B120	121.8	123.0
B128	129.8	131.0
B136	137.8	139.0
B144	145.8	147.0
B158	159.8	161.0
B173	174.8	176.0
B180	181.8	183.0
B195	196.8	198.0
B210	211.8	213.0
B240	240.3	241.5
B270	270.3	271.5
B300	300.3	301.5

C BELTS		
BELT # Standard	Pitch Length	Outside Length
C51	53.9	55.0
C60	62.9	64.0
C68	70.9	81.0
C75	77.9	79.0
C81	83.9	85.0
C85	87.9	89.0
C90	92.9	94.0
C96	98.9	100.0
C105	107.9	109.0
C112	114.9	116.0
C120	122.9	124.0
C128	130.9	132.0
C136	138.9	140.0
C144	146.9	148.0
C158	160.9	162.0
C162	164.9	166.0
C173	175.9	177.0
C180	182.9	184.0
C195	197.9	199.0
C210	212.9	214.0
C240	240.9	242.0
C270	270.9	272.0
C300	300.9	302.0
C360	360.9	362.0
C390	390.9	392.0
C420	420.9	422.0

D BELTS		
BELT # Standard	Pitch Length	Outside Length
D120	123.3	125.0
D128	131.3	133.0
D144	147.3	149.0
D158	1613	163.0
D162	165.3	167.0
D173	176.3	178.0
D180	183.3	185.0
D195	198.3	200.0
D210	213.3	215.0
D240	240.8	242.0
D270	270.8	272.5
D300	300.8	302.5
D330	330.8	332.5
D360	360.8	362.5
D390	390.8	392.5
D420	420.8	422.5
D480	480.8	482.5
D540	540.8	542.5
D600	600.8	602.5

3V BELTS		5V BELTS		8V BELTS	
Belt No.	Belt Length	Belt No.	Belt Length	Belt No.	Belt Length
3V250	25.0	5V500	50.0	8V1000	100.0
3V265	26.5	5V530	53.0	8V1060	106.0
3V280	28.0	5V560	56.0	8V1120	112.0
3V300	30.0	5V600	60.0	8V1180	118.0
3V315	31.5	5V630	63.0	8V1250	125.0
3V335	33.5	5V670	67.0	8V1320	132.0
3V355	35.5	5V710	71.0	8V1400	140.0
3V375	37.5	5V750	75.0	8V1500	150.0
3V400	40.0	5V800	80.0	8V1600	160.0

E BELTS		
BELT # Standard	Pitch Length	Outside Length
E180	184.5	187.5
E195	199.5	202.5
E210	214.5	217.5
E240	241.0	244.0
E270	271.0	274.0
E300	301.0	304.0
E330	331.0	334.0
E360	361.0	364.0
E390	391.0	394.0
E420	421.0	424.0
E480	481.0	484.0
E540	541.0	544.0
E600	601.0	604.0

3V BELTS		5V BELTS		8V BELTS	
3V425	42.5	5V850	85.0	8V1700	170.0
3V450	45.0	5V900	90.0	8V1800	180.0
3V475	47.5	5V950	95.0	8V1900	190.0
3V500	50.0	5V1000	100.0	8V2000	200.0
3V530	53.0	5V1060	106.0	8V2120	212.0
3V560	56.0	5V1120	112.0	8V2240	224.0
3V600	60.0	5V1180	118.0	8V2360	236.0
3V630	63.0	5V1250	125.0	8V2500	250.0
3V670	67.0	5V1320	132.0	8V2650	265.0
3V710	71.0	5V1400	140.0	8V2800	280.0
3V750	75.0	5V1500	150.0	8V3000	300.0
3V800	80.0	5V1600	160.0	8V3150	315.0
3V850	85.0	5V1700	170.0	8V3350	335.0
3V900	90.0	5V1800	180.0	8V3550	355.0
3V950	95.0	5V1900	190.0	8V3750	375.0
3V1000	100.0	5V2000	200.0	8V4000	400.0
3V1060	106.0	5V2120	212.0	8V4250	425.0
3V1120	112.0	5V2240	224.0	8V4500	450.0
3V1180	118.0	5V2360	236.0	*8V5000	500.0
3V1250	128.0	5V2500	250.0		
3V1320	132.0	5V2650	265.0		
3V1400	140.0	5V2800	280.0		
		5V3000	300.0		
		5V3150	315.0		
		5V3350	335.0		
		5V3550	355.0		

For example, if the 60-inch "B" section belt shown is manufactured 3/10 of an inch longer, it will be code marked 53 rather than the 50 shown. If made 3/10 of an inch shorter, it will be code marked 47. While both of these belts have the belt number B60 they cannot be used satisfactorily in a set because of the difference in their actual length.

TYPICAL CODE MARKING

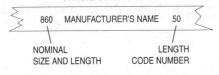

860   MANUFACTURER'S NAME   50

NOMINAL
SIZE AND LENGTH

LENGTH
CODE NUMBER

MELTING POINT AND RELATIVE CONDUCTIVITY OF DIFFERENT METALS AND ALLOYS

Metals	Relative Conductivity	Melting Point °F
Pure silver	106.0	1760
Pure copper	100.0	1980
Refined and crystalized copper	99.9	——
Telegraphic silicious bronze	98.0	——
Alloy of copper and silver (50%)	86.65	——
Silicide of copper, 4% Si	75.0	——
Pure gold	71.3	1950
Pure aluminum	64.5	1220
Silicide of copper, 12% Si	54.7	——
Tin with 12% of sodium	46.9	——
Telephonic silicious bronze	35.0	——
Copper with 10% of lead	30.0	——
Pure zinc	29.2	790
Telephonic phosphor-bronze	29.0	——
Silicious brass, 25% zinc	26.4	——
Brass with 35% zinc	21.59	——
Phosphor-tin	17.7	——
Alloy of gold and silver (50%)	16.12	——
Swedish iron	16.4	2800
Pure platinum	16.3	3230
Pure banca tin	15.5	442
Antimonial copper	12.7	——
Aluminum bronze (10%)	12.6	——
Siemens Steel	12.0	——
Copper with 10% of nickel	10.6	——
Cadmium amalgam (15%)	10.2	——
Dronier mercurial bronze	10.14	——
Arsenical copper (10%)	9.1	——
Bronze with 20% tin	8.4	——
Pure lead	7.8	620
Phosphor-bronze, 10% tin	6.5	——
Pure nickel	5.8	2640
Phosphor-copper, 9% phosphor	4.9	——
Antimony	4.4	1167

CHAPTER 12
GLOSSARY

10Base2 - 10 Mbps, baseband, in 185 meter segments. The IEEE 802.3 substandard for ThinWire, coaxial, Ethernet.

10BaseT - 10 Mbps, baseband, over twisted pair. The IEEE 802.3 substandard for unshielded twisted pair Ethernet.

110 Type Block - A wire connecting block that terminates 100 to 300 pairs of wire.

66-Type Block - A type of wire connecting block that is used for twisted pair cabling cross connections. It holds 25 pairs in up to four columns.

AC Current- Electrical current which reverses direction at regular intervals (cycles) due to a change in voltage which occurs at the same frequency.

AC Line Filter - Absorbs electrical interference from motors, etc.

AC Sine Wave - A symmetrical waveform that contains 360 electrical degrees.

AC Voltage - Electrical pressure that reverses direction at regular intervals called cycles.

Absorption - Loss of power in an optical fiber, resulting from conversion of optical power into heat and caused principally by impurities, such as transition metals and hydroxyl ions.

Access Line - A line or circuit that connects a customer site to a network switching center or local exchange. Also known as the local loop.

Aerial Cable - Telecommunications cable installed on aerial supporting structures such as poles, sides of buildings, and other structures.

Alternator - Converts mechanical energy into electrical energy.

Ambient Temperature - The temperature of the surroundings.

American National Standards Institute (ANSI) - A private organization that coordinates some US standards setting.

American Standard Code for Information Interchange (ASCII) - A standard character set that (typically) assigns a 7-bit sequence to each letter, number, and selected control character.

American Wire Gauge (AWG) - Standard used to describe the size of a wire. The larger the AWG number, the smaller (thinner) the described wire.

Ammeter - An instrument (meter) for measuring electrical current.

Ampacity - The amount of current (measured in amperes) that a conductor can carry without overheating.

Ampere (A) - Unit of current measurement. The amount of current that will flow through a one ohm resistor when one volt is applied.

Ampere-hour - The quantity of electricity equal to the flow of a current of one ampere for one hour.

Amplitude - The size, in voltage, of signals in a data transmission.

Angular Misalignment - The loss of optical power caused by deviation from optimum alignment of fiber to fiber or fiber to waveguide.

Annunciator - A sound generating device that intercepts and speaks the condition of circuits or circuits' operations.

Anode - The positive electrode in an electrochemical cell (battery) toward which current flows.

Apparent Power (P$_A$) - Product of the voltage and current in a circuit calculated without considering the phase shift that might be present between the voltage and the current. Expressed in terms of Volt - Ampere (VA).

Approved Ground - A grounding bus or strap in a building that is suitable for connecting to data communication equipment.

Arcing - A luminous discharge formed by the span of electrical current across a space between electrodes or conductors.

Arc Tube - The light-producing element of an HID lamp.

Armature - The rotating part of a DC motor.

Arrestor (lightning) - A device that reduces the voltage of a surge applied to its terminals and restores itself to its original operating condition.

Attenuation - Denotes the loss in strength of power between that transmitted and that received. Usually expressed as a ratio in dB (decibel).

Attenuator - A device that reduces signal power in a fiber optic link by inducing loss.

Autotransformer - Changes voltage level using the same common coil for both the primary and the secondary.

Avalanche Current - Current passed when a diode malfunctions.

Average Value (Vavg) - The mathematical mean of all instantaneous voltage values in a sine wave.

B - Byte.

b - bit.

Back Electromagnetic Force - The voltage created in an inductive circuit by a changing current flowing through the circuit.

Back Reflection, Optical Return Loss - Light reflected from the polished end of a fiber caused by the difference of refractive indices.

Backboard - A wooden (or metal) panel used for mounting equipment usually on a wall.

Backbone - The main connectivity device of a distributed system. All systems that have connectivity to the backbone will connect to each other.

Ballast - An electrical circuit component used with fluorescent lamps to provide the voltage necessary to strike the mercury arc within the lamp, and then to limit the amount of current that flows through the lamp.

Bandwidth - Technically, the difference, in Hertz (Hz), between the highest and lowest frequencies of a transmission channel.

Bank - An assemblage of fixed contacts.

Battery - A device that converts chemical energy into electrical current.

Baud - A unit of signaling speed. The speed in Baud is the number of discrete conditions or signal elements per second.

Bidirectional Current - Has both positive and negative values.

Bits Per Second (bps) - Basic unit of measurement for serial data transmission capacity, abbreviated as k bps, or kilobit/s, for thousands of bits per second; m bps, or megabit/s, for millions of bits per second; g bits, or gigabit/s for billions of bits per second.

Blocking Diode - A diode used to prevent current flow in a photovoltaic array during times when the array is not producing electricity.

Bonding - A very low impedance path accomplished by permanently joining non-current-carrying metal parts. It is done to provide electrical continuity and to offer the capacity to safely conduct any current.

Bonding Jumper - A conductor used to assure the required electrical connection between metal parts of an electrical system.

Bonding Conductor - The conductor that connects the non-current-carrying parts of electrical equipment, cable raceways, or other enclosures to the approved system ground conductor.

Bond Wire - Bare grounding wire that runs inside of an armored cable.

Branch Circuit - Conductors between the last overcurrent device and the outlets.

Branch Circuit, Multiwire - A branch circuit having two or more ungrounded circuit conductors, each having a voltage difference between them, and a grounded circuit conductor (neutral) having an equal voltage difference between it and each ungrounded conductor.

Break - The number of separate places on a contact that open or close a circuit.

Breakout Box - A device that allows access to individual points on a physical interface connector for testing and monitoring.

Breakout Cable - A multifiber cable where each fiber is protected by an additional jacket and strength element beyond the overall cable.

British Thermal Unit (BTU) - The amount of heat necessary to raise the temperature of one pound of water 1°F.

Brushes - Sliding contacts that provide the connection between the external power circuit and the commutator of a DC motor.

Bus - A group of conductors that serve as a common connection for circuits.

Bus Bar - The heavy copper or aluminum bar used to carry currents in switchboards.

Busway - A metal enclosed distribution of bus bars.

Cable - One or more insulated or non-insulated wires used to conduct electrical current.

Calorie (Cal) - The amount of heat required to raise 1 gallon of water 1°C.

Capacitance (C) - The ability of a circuit or component to store an electrical charge and is measured in farads (F).

Capacitive Circuit - A circuit in which current leads voltage (voltage lags current).

Capacitive Reactance (Xc) - The opposition to current flow by a capacitor measured in Ohms (Ω).

Capacitor - A device that stores electrical energy by means of an electrostatic field.

Cathode - The negative electrode in an electrochemical cell (Battery).

Choke Coil - An inductor used to limit the flow of AC.

Circuit - A complete path through which electricity may flow.

Circuit Breaker - A device used to open and close a circuit by automatic means when a predetermined level of current flows through it.

Circular Mil (cm) - A measurement of the cross-sectional area of a conductor. kcmil equals 1000 circular mils.

Closed Circuit - A continuous path providing for electrical flow.

Coil - A winding of insulated conductors arranged to produce magnetic flux.

Common Noise - Noise produced between the ground and the hot or the ground and the neutral lines.

Commutator - A series of copper segments connected to the armature of a DC motor.

Conductance - The measure of the ability of a component to conduct electricity expressed in mohs.

Conductor - A substance which offers little resistance to the flow of electrical currents. Insulated copper wire is the most common.

Conduit Body - The part of a conduit system, at the junction of two or more sections of the system, that allows access through a removable cover. Most commonly known as condulets, LBs, LLs, and LRs.

Contacts - The conducting part of a switch that operates with another conducting part to make or break a circuit.

Contactor - A control device that uses a small current to energize or de-energize the load connected to it.

Continuous Load - A load whose maximum current continues for three hours or more.

Core - The central, light-carrying part of an optical fiber.

Coulomb - An electrical current of 1 ampere per second.

Crosstalk - The unwanted energy transferred from one circuit or wire to another which interferes with the desired signal. Usually caused by excessive inductance in a circuit.

Current - The flow of electricity in a circuit, measured in amperes.

Cut-out Box - A surface mounted electrical enclosure with a hinged door.

Cutback Method - A technique for measuring the loss of bare fiber by measuring the optical power transmitted through a long length then cutting back to the source and measuring the initial coupled power.

Cycle - Measured in hertz (Hz), it is the flow of AC in one direction and then in the opposite direction in one time interval completing one wavelength. Hertz measures cycles per second.

Daisy Chaining - The connection of multiple devices in a serial fashion.

Data Rate - The number of bits of information in a transmission system, expressed in bits per second (bps), and which may or may not be equal to the signal or baud rate.

dB - Decibel referenced to a microwatt.

dBm - Decibel referenced to a milliwatt.

Decibel - 1) A standard logarithmic unit for the ratio of two powers, voltages, or currents. In fiber optics, the ratio is power. 2) Unit for measuring relative strength of a signal parameter such as power or voltage.

Deep Cycle - Battery type that can be discharged to a large fraction of capacity. See Depth of Discharge.

Degauss - To remove residual permanent magnetism.

Delta Connection - A connection that has each coil end connected end-to-end to form a closed loop.

Depth of Discharge (DOD) - The percent of the rated battery capacity that has been withdrawn.

Device (Also used as wiring device) - The part of an electrical system that is designed to carry, but not use, electrical energy.

Diac - A thyristor that triggers in either direction when its breaker voltage is exceeded.

Diode - Electronic component that allows current flow in one direction only.

Direct Current (DC) - Electrical current which flows in one direction only.

DC Compound Motor - The field is connected in both series and shunt with the armature.

DC Permanent Magnet Motor - Uses magnets, not a coil, for the field winding.

DC Series Motor - The field is connected in series with the armature.

DC Shunt Motor - The field is connected in parallel with the armature.

DC Voltage - Voltage that flows in one direction only

Double Break Contacts - Contacts that break the current in two separate places.

Dispersion - Causes a spreading of light as it propagates through an optical fiber. Three types are modal, material, and waveguide.

Dual Voltage Motor - A motor that operates at more than one voltage level.

Eddy Current - Unwanted current induced in the metal field structure of a motor.

Effective Current - The value of AC which causes the same heating effect as a given value of DC. For sine wave AC, the effective current is 0.7071.

Effective Value - Also called the Root-Mean-Square (RMS), it produces the same I^2R power as an equal DC value.

Efficiency (EFF) - The ratio of output power to input power, usually expressed as a percentage.

Electricity - The movement of electrons through a conductor.

Electromagnet - Coil of wire that exhibits magnetic properties when current passes through it.

Electron - The subatomic unit of negative electricity expressed as a charge of 1.6×10^{-19} coulomb.

Electronics - The science of treating charge flow in vacuums, gases and crystal lattices.

Energy - The capacity to do work.

Equalization - The process of restoring all cells in a battery to an equal state of charge.

Ethernet - A 10-Mbps, coaxial standard for LANs. All nodes connect to the cable where they contend for access via CSMA/CD.

Excess Loss - In a fiber-optic coupler, the optical loss from that portion of light that does not emerge from the nominally operational ports.

Excitation - The power required to energize the magnetic field of transformers, generators and motors.

Extrinsic Loss - In a fiber interconnection, that portion of loss that is not intrinsic to the fiber but is related to imperfect joining, which may be caused by the connector or splice.

Farad (F) - The unit of measurement of capacitance.

Fault Current - Any current that travels an unwanted path, other than the normal operating path of an electrical system.

Feeder - Circuit conductors between the service and the final branch circuit overcurrent device.

Ferrule - A precision tube that holds a fiber for alignment for interconnection or termination.

Fiber Optics - A technology that uses light as a digital information carrier.

Field - The stationary windings (magnets) of a DC motor.

Filament - A conductor that has a high enough resistance to cause heat.

Filter - A combination of circuit elements which is specifically designed to pass certain frequencies and resist all others.

Flashover - A disruptive electrical discharge around or over (but not through) an insulator.

Fluorescence - The emission of light by a substance when exposed to radiation or the impact of particles.

Flux - An electrical filled energy distributed in space and represented diagrammatically by means of flux lines denoting magnetic or electrical forces.

Footcandle (fc) - The amount of light produced by a lamp measured in lumens divided by the area that is illuminated.

Four Wire Circuits - Telephone circuits which use two separate one-way transmission paths of two wires each, as opposed to regular local lines which usually only have two wires to carry conversations in both directions.

Frequency - The number of times per second a signal regenerates itself at a peak amplitude. It can be expressed in hertz (Hz), kilohertz (kHz), megahertz (MHz), etc.

Full-Load Current (FLC) - The current required by a motor to produce the full-load torque at the motor's rated speed.

Full-Load Torque (FLT) - The torque required to produce the rated power of the motor at full speed.

Fuse - A protective device, also called an OCPD, with a fusible element that opens the circuit by melting when subjected to excessive current.

Gain - A ratio of the amplitude of the output signal to the input signal.

Gap Loss - Loss resulting from the end separation of two axially aligned fibers.

Gb - Gigabit. One billion bits of information.

Gbyte - Gigabyte. One billion bytes of data.

Ghost Voltage - A voltage that appears on a motor that is not connected.

Grid - Term used to describe an electrical utility distribution network.

Ground - An electrical connect (on purpose or accidental) between an item of equipment and the earth.

Grounding - The connection of all exposed non-current carrying metal parts to the earth.

Ground Fault - A condition in which current from a hot power line is flowing to the ground.

Guy - A wire having one end secured and the other fastened to a pole or structure under tension.

Ground Fault Circuit Interrupter (GFCI) - An electrical device which protects personnel by detecting hazardous ground faults and quickly disconnects power from the circuit.

Harmonic - 1) A sinusoid which has a frequency which is an integral multiple of a certain frequency. 2) The full multiple of a base frequency.

Heater - A device that is placed in a motor starter to measure the amount of current in the power line.

Heat Sink - A piece of metal used to dissipate the heat of solid-state components mounted on it.

Heating Element - A conductor (wire) that offers enough resistance to produce heat when connected to power.

Henry (H) - The unit of measure of inductance in which a current changing its rate of flow one ampere per second induces an electromotive force of one volt.

Hertz (Hz) - The unit of measure of the frequency of the AC sine wave to complete a cycle. One Hertz is equal to one cycle of the AC sine-wave per second.

HID Lamp - High Intensity Discharge Lamp.

Holding Current - The minimum current necessary for an SRC to continue conducting.

Horsepower (HP) - A unit of power equal to 746 watts that describes the output of electric motors.

Hub - A device which connects to several other devices usually in a star topology. Also called: concentrator, multiport repeater or multi-station access unit (MAU).

Hybrid - An electronic circuit that uses different cable types to complete the circuit between systems.

Impedance (Z) - Is the total opposition offered to the flow of AC from any combination of resistance, inductive reactance and capacitive reactance and is measured in Ohms (Ω).

Inductance (L) - The property of a circuit that determines how much voltage will be induced into it by a change in current of another circuit and is measured in henrys (H).

Inductive Circuit - A circuit in which current lags voltage.

Inductive Reactance (X_L) - The opposition to the flow of AC in a circuit due to inductance and is measured in Ohms (Ω).

In-Phase - The state when voltage and current reach their maximum amplitude and zero level at the same time in a cycle.

Insertion loss - The loss of power that results from inserting a component, such as a connector or splice, into a previously continuous path.

Insulator - Material that current cannot flow through easily.

Integrated Circuit or IC - A circuit in which devices such as transistors, capacitors, and resistors are made from a single piece of material and connected to form a circuit.

Interface - The point that two systems, with different characteristics, connect.

Isolated Grounded Receptacle - Minimizes electrical noise by providing a separate grounding path.

Isolation Transformer - A one to one transformer that is used to isolate the equipment at the secondary from earth ground.

Jack - A receptacle (female) used with a plug (male) to make a connection to in-wall communications cabling or to a patch panel.

Jacket - The protective and insulating outer housing on a cable. Also called a sheath.

Jogging - The frequent starting and stopping of a motor.

Joule - A unit of electrical energy also called a watt-second; the transfer of one watt for one second.

Jumper - Patch cable or wire used to establish a circuit, often temporarily, for testing or diagnostics.

Junction Box - A box, usually metal, that encloses cable connections.

KB - Kilobyte. One thousand bytes.

Kb - Kilobit. One thousand bits.

Kbps - Kilobits per second. Thousand bits per second.

kcmil - One thousand circular mils.

kVA - Kilovolt-amperes (1,000 volt amps)

kVAR - Kilovar (1,000 reactive volt amps)

kV - Kilovolt (1,000 volts)

kW - Kilowatt (1,000 watts)

kWH - Kilowatt hour - The basic unit of electrical energy for utilities equal to a thousand watts of power supplied for one hour.

Lamp - A light source. Reference is to a light bulb, rather than a lamp.

Leakage Current - Current that flows through insulation.

Leg (circuit) - One of the conductors in a supply circuit in which the maximum voltage is maintained.

Load - The amount of electric power used by any electrical unit or appliance at any given moment.

Location, Damp - Partially protected locations, such as under canopies, roofed open porches, etc. Also, interior locations that are subject only to moderate degrees of moisture, such as basements, barns, etc.

Location, Wet - Locations underground, in concrete slabs, where saturation occurs, or outdoors.

Locked Rotor - Condition when a motor is loaded so heavily that the shaft cannot turn.

Locked Rotor Current (LRC) - The steady-state current taken from the power line with the rotor locked and the voltage applied.

Locked Rotor Torque (LRT) - The torque a motor produces when the rotor is stationary and full power is applied.

Loss Budget - The amount of power lost in a fiber optic link. Used in terms of the maximum amount of loss that can be tolerated by a given link.

Loss, optical - The amount of optical power lost as light is transmitted through fiber, splices, couplers, and the like.

Lumen (lm) - The unit used to measure the total amount of light produced by a light source.

Mbps - Million bits per second.

Mbyte - Megabyte. Million bytes of information.

Magnetic Field - The invisible field produced by a current-carrying conductor or coil, permanent magnet, or the earth itself, that develops a north and south polarity.

Magnetic Flux - The invisible lines of force that make up the magnetic field.

Magnetic Induction - The setting up of magnetic flux lines in a material by an electric current. The number of lines is measured in maxwells.

Maxwell - The unit of measurement of the total number of magnetic flux lines in a magnetic field.

Mechanical Splice - A semipermanent connection between two fibers made with an alignment device and index matching fluid or adhesive.

Micron (m) - A unit of measure, 10^{-6} m, used to measure wavelength of light.

Mode - A single electromagnetic field pattern that travels in fiber.

Motor - A machine that develops torque (rotating mechanical force) on a shaft to produce work.

Motor Efficiency - The effectiveness of a motor to convert electrical energy into mechanical energy.

Motor Starter - An electrically operated switch (contactor) that includes overload protection.

Motor Torque - The force that produces rotation in the shaft.

Multiplex - To combine multiple input signals into one for transmission over a single high-speed channel. Two methods are used: (1) frequency division, and (2) time division.

Mutual Inductance - The effect of one coil inducing a voltage into another coil.

Nanometer (nm) - A unit of measure, 10^{-9} m, used to measure the wavelength of light.

NEC - National Electrical Code, which contains safety rules for all types of electrical installations.

No-Load Current - The current demand of a transformer primary when no current demand is made on the secondary.

Normally Open Contacts - Contacts that are open before being energized.

Normally Closed Contacts - Contacts that are closed before being energized.

Ohm - The unit of measurement of electrical resistance. One ohm of resistance will allow one ampere of current to flow through a pressure of one volt.

Ohm's Law - A law which describes the mathematical relationship between voltage, current and resistance.

Open Circuit - A condition that provides no path for electric current to flow in a circuit.

Open Circuit Voltage - The maximum voltage produced by a photovoltaic cell, module, or array without a load applied.

Optical Power - The amount of radiant energy per unit of time, expressed in linear units of watts or on a logarithmic scale, in dBm (where 0 dB = 1 mW) or dB (where 0 dB = 1 W).

Oscillation - Fluctuations in a circuit.

Out of Phase - Having AC sine waveforms that are of the same frequency, but are not passing through corresponding values at the same instant.

Outlet - The place in the wiring system where the current is taken to supply equipment.

Overcurrent - Too much current.

Overload Protection - A device that prevents overloading a circuit or motor such as a fuse or circuit breaker.

Parallel Circuit - Contains two or more loads and has more than one path through which current flows.

Peak Value - The maximum value of either the positive or negative alteration of a sine wave.

Period - The time required to produce one cycle of a waveform.

Phase - The fractional part of a period through which the time variable of a periodic quantity has moved.

Phase Converter - A device that derives three phase power from single phase power.

Phase Shift - The state when voltage and current in a circuit do not reach their maximum amplitude and zero level at the same time.

Photovoltaic - Changing light into electricity.

Pigtail - A short length of fiber permanently attached to a component.

Polarity - The particular state of an object, either positive or negative, which refers to the two electrical poles, north, and south.

Power (Watts) - A basic unit of electrical energy, measured in watts.

Power Budget - The difference (in dB) between the transmitted optical power (in dBm) and the receiver sensitivity (in dBm).

Power Factor - The ratio of true power (kW) to Apparent power (kVA) for any given load and period of time.

Primary Winding - The coil of a transformer which is energized from a source of alternating voltage and current. The input side.